MILITARY AND VETERAN ISSUES

# U.S. GEOLOGICAL SURVEY ROADMAP FOR UNMANNED AIRCRAFT SYSTEMS

# MILITARY AND VETERAN ISSUES

Additional books in this series can be found on Nova's website under the Series tab.

Additional e-books in this series can be found on Nova's website under the e-book tab.

MILITARY AND VETERAN ISSUES

# U.S. GEOLOGICAL SURVEY ROADMAP FOR UNMANNED AIRCRAFT SYSTEMS

KATHRYN MILLER
EDITOR

nova publishers
*New York*

**Library of Congress Cataloging-in-Publication Data**

ISBN: 978-1-63483-566-4

*Published by Nova Science Publishers, Inc. † New York*

# CONTENTS

# PREFACE

The U.S. Department of the Interior (DOI) is responsible for protecting the natural resources and heritage contained on almost 20 percent of the land in the United States. This responsibility requires acquisition of remotely sensed data throughout vast lands, including areas that are remote and potentially dangerous to access. One promising new technology for data collection is unmanned aircraft systems (UAS), which may be better suited (achieving superior science, safety, and savings) than traditional methods. UAS, regardless of their size, have the same operational components: aircraft, payloads, communications unit, and operator control unit. This book provides operational procedures and lessons learned from completed proof-of-concept UAS missions in areas such as wildlife management, resource monitoring, and public land inspections. This information provides not only an implementation framework but can also help increase the awareness by resource managers, scientists, and others of the ability of UAS technology to advance data quality, improve personnel safety, and reduce data acquisition costs.

In: U.S. Geological Survey Roadmap …
Editor: Kathryn Miller
ISBN: 978-1-63483-566-4


*Chapter 1*

# U.S. GEOLOGICAL SURVEY UNMANNED AIRCRAFT SYSTEMS (UAS) ROADMAP 2014*

***Jill Cress, Michael Hutt, Jeff Sloan, Mark Bauer, Mark Feller and Susan Goplen***

## SUMMARY

The Department of the Interior (DOI) is responsible for managing more than 500 million acres of surface land, or about one-fifth of the land in the United States. Meeting the challenges of various mandated missions of the Department is complex, challenging, and rewarding. Since its inception in 2008, the U.S. Geological Survey (USGS) National Unmanned Aircraft Systems (UAS) Project Office has been working in collaboration across the DOI in technology transfer, applications development, and the implementation of UAS technology. UAS are proving to be a cost-effective, efficient capability that enhances our ability to analyze the effects of climate change, respond to natural hazards, study rates and consequences of landscape change, conduct wildlife inventories, and support land management missions.

Since the initial United States Geological Survey Unmanned Aircraft Systems Roadmap 2010– 2025 was released on July 29, 2011, interest in the utilization of UAS has continued to progress at an exponential rate. This

---

* This is an edited, reformatted and augmented version of U.S. Geological Survey Open-File Report 2015–1032, issued March 2015.

progress can be partially attributed to the easier acceptance of new technologies by USGS scientists who were exposed to computer technology at an early age and are constantly looking for innovative ways to adapt newer technology to do their jobs better, more effectively, and more safely. They are willing to question the traditional paradigms and methods to evaluate if observing the Earth in a different way might provide new insights, enhanced perspective, and new solutions.

A snapshot of the global UAS community in 2014 shows more than 175 U.S. universities involved with UAS-related programs, 70 countries actively involved with some aspect of UAS manufacturing, 200 UAS-related manufacturers in the United States, 715 global UAS-related manufacturers and nearly 2,500 UAS platforms on the market. Key contributors to the growth of UAS technology were several milestones specific to UAS established by the U.S. Congress in the Federal Aviation Administration (FAA) Modernization and Reform Act of 2012 (Public Law 112-95, 126 Stat. 11). The act established a 2015 deadline for integration of UAS into the National Airspace System (NAS), six UAS test sites, and support for UAS operations in the Arctic. UAS, after they are integrated into the NAS, will provide more than 70,000 U.S. jobs and add more than $13 billion to the U.S. economy (Association for Unmanned Vehicle Systems International, 2013; Teal Group Corporation, 2013).

The DOI USGS use of UAS technology dates back to 2004 when a UAS was used to acquire data during a volcanic event on Mount Saint Helens, Washington. After carefully monitoring the rapid advancement of UAS, the USGS determined that UAS were ready to be employed for scientific, environmental, and land-management applications. The USGS National UAS Project Office was created in May 2008 to orchestrate an effective evaluation and transfer of UAS technology into the DOI's decisionmaking toolbox. USGS is only part of an energized and collaborative team working across the DOI, FAA, National Aeronautics and Space Administration (NASA), National Oceanic and Atmospheric Administration, U.S. Forest Service, academia, and private industry. The DOI Aviation Executive Committee also formed a UAS working group to develop a cohesive, long-range, crosscutting UAS strategy, and the DOI Remote Sensing Working Group is maintaining an awareness of UAS capabilities.

Similar to the use of the Internet, Geographic Information Systems, and Global Positioning Systems, use of UAS are enabling us to be better stewards of the land. UAS technology provides scientists a way to look longer, closer, and more frequently at some of the most remote areas of the Earth, places that were previously too dangerous or too expensive to monitor in detail. The flexibility of operations and relative low cost of small UAS (sUAS) enhance our ability to track long-term landscape and environmental change. In addition, we can quickly assess landscape-altering events, such as wildfires, floods, and volcanoes.

The DOI requirements for UAS missions are "bubbling-up" from field-level staffs. The innovation and dedication of the DOI scientists and resource managers are readily apparent as they turn to UAS to perform their old jobs more cost effectively and safely and to do new jobs that hadn't been previously possible. Wildlife biologists first implemented the technology (monitoring and inventorying wildlife), followed by geologists (detecting landslides, mapping fault zones), hydrologists (monitoring shoreline erosion and stream temperature gradients) and ecologists (habitat mapping). The public safety components of the DOI are very interested in using UAS to support search and rescue, monitoring pipelines, and wildland firefighting. In general, if something can be observed with a manned aircraft today, we anticipate observing it with an unmanned system in the future. Our goal is to achieve a 10-to-1 cost savings by using a sUAS as compared with traditional manned aircraft for small projects.

We have taken the approach that smaller systems (less than 55 pounds) will be the first UAS approved by the FAA for routine use in the NAS. We are currently using the AeroVironment Raven and Honeywell T-Hawk to conduct our operational test and evaluation of sUAS technology. The most exciting UAS-related development is the miniaturization and variety of readily available sensor packages. Using UAS, we are able to tailor sensor solutions to meet project requirements. We can readily obtain high-resolution video, acquire thermal imagery, collect point cloud data, and generate high-resolution digital elevation models at a fraction of the cost of conventional surveying methods.

Because the DOI is concerned with expenses related to system acquisition, maintenance, and operator training, we are defining crosscutting requirements that minimize the number of systems we operate and yet maximize their use. We also recognize that larger UAS will play a role in meeting our mission requirements. We anticipate contracting for data services with commercial UAS vendors in the future to meet long-duration or specialized acquisition requirements (such as State or national aerial photography or lidar surveys).

UAS technology will allow us to do more with less and in the process enhance our ability to provide unbiased scientific information to better enable informed decisions. We fully expect that by 2020 UAS will emerge as the primary platform for all DOI aerial remote sensing applications. In summary, compared to traditional data acquisition methods, UAS data acquisition can be more

- Economical
- Safe
- Efficient.

UAS technology allows us to do things we couldn't do before, such as

- Enhanced observations
- New science
- Better information to support informed decisions.

UAS technology

- Will supplement, not replace, other observation techniques
- Will emerge as the primary platform for DOI remote sensing applications.

I am grateful to the USGS National UAS Project Office (Jeff Sloan, Jill Cress, Susan Goplen, Mark Feller, Mark Bauer, Isaac Anderson), Mark Bathrick and his outstanding staff at the Office of Aviation Services, Lance Brady (Bureau of Land Management), Doug Clark (Bureau of Reclamation), Jim Traub (National Park Service), David Johncox, Mathew Larsen, Tim Newman, Bruce Quirk and their staffs at USGS and, most important, the numerous field scientists and resource managers throughout the DOI for their support of the USGS National UAS Project Office and for helping us introduce an innovative, paradigm-shifting technology.

Michael E. Hutt (retired)
USGS National Unmanned Aircraft Systems Project Office

## INTRODUCTION

The U.S. Department of the Interior (DOI) is responsible for protecting the natural resources and heritage contained on almost 20 percent of the land in the United States. This responsibility requires acquisition of remotely sensed data throughout vast lands, including areas that are remote and potentially dangerous to access. The challenges associated with meeting this requirement for scientific data and information about our public lands, their ecosystems, and their natural inhabitants have produced many innovative remote sensing strategies. Although DOI scientists can obtain high accuracy in-place data based on ground measurements and sample collections, the scale of public lands makes this approach unrealistic for a large area or periodic survey. Sensors aboard piloted aircraft can, and frequently do, cover large areas with high-resolution imagery, but increasing costs and potential safety issues can limit their availability. The fact that many of these manned aircraft flights must take place in remote or dangerous areas, such as the North Slope of Alaska or the Everglades, only increases these concerns. Although satellite imagery (through the Landsat and other data-collection missions) has proven to be a vital resource, it can be limited by resolution and orbital dependencies. In response to these ongoing data-collection challenges it is crucial for DOI to continually evaluate new technologies in order to find safer, lower cost, and more flexible methods for collecting remotely sensed data (fig. 1).

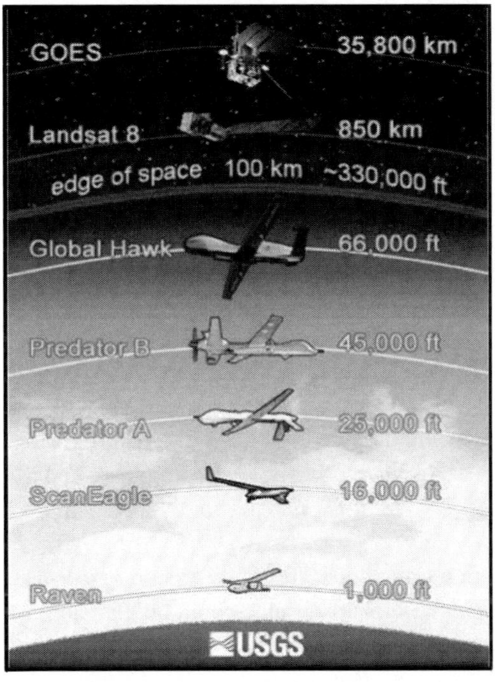


Figure 1. Illustration showing operational altitudes of remote sensing platforms.

One promising new technology for data-collection is unmanned aircraft systems (UAS), which may be better suited (achieving superior science, safety, and savings) than traditional methods. This Roadmap provides operational procedures and lessons learned from completed proof-of-concept UAS missions in areas such as wildlife management, resource monitoring, and public land inspections. This information provides not only an implementation framework but can also help increase the awareness by resource managers, scientists, and others of the ability of UAS technology to advance data quality, improve personnel safety, and reduce data acquisition costs.

## Background

Unmanned aircraft systems (UAS) continue to be the subject of spirited debate about their basic operations, reliability, and safety. As the systems have been more widely accepted in military applications, questions about UAS compatibility with existing air and ground systems have also been hotly debated. But clearly, operations of UAS within military missions and military operational airspace have become essential, making the use of UAS in support of civil agency missions the next logical step.

The Department of the Interior (DOI) first used UAS technology in 2004 to acquire data during a volcanic event on Mount Saint Helens, Washington. This successful mission demonstrated that UAS technology had matured enough that it could become a cost-effective remote sensing tool for scientific, environmental, and land management applications. In 2005, the U.S. Geological Survey (USGS) Land Remote Sensing Program sponsored an emerging technology investigation of UAS technology that concluded the following:

- military use of UAS technology was increasing at an exponential rate,
- many civil agencies were establishing UAS program offices,
- U.S. universities were offering UAS programs, and
- the number of UAS vendors was rapidly increasing.

This investigation concluded that UAS technology had transitioned from military use only to a more mainstream technology that could be used as a new remote sensing tool for the DOI.

As a next step towards operational use of UAS technology in the DOI, the USGS created the National UAS Project Office (NUPO) in the spring of 2008.

The NUPO was designed to lead and coordinate efforts to implement UAS technology as a cost-effective option for acquiring remotely sensed data within the DOI. Results from these efforts, including current operational status within the DOI and information on completed proof-of-concept missions, are fully documented within this Roadmap.

## Characteristics of Unmanned Aircraft Systems

Any discussion of operational implementation of unmanned aircraft systems (UAS) should start with a clear understanding of the underlying technology—technology that has rapidly progressed from its simple application as a weekend hobbyist's prototype to today's increasing variety of highly automated, tested, and operational systems. Although today's UAS come in an increasingly wide array of shapes and sizes, they all share similar characteristics. To start, each UAS has three operational phases: launch, flight, and recovery. Methods of launching fixed-wing aircraft include traditional hard-surface takeoff, hand thrown, pneumatic catapult, and bungee, whereas all rotorcraft use vertical take-off and landing. Flight can be accomplished by either preloading global positioning systems waypoint navigation flight lines into the onboard flight control system or remote control by a pilot on the ground. The last phase is recovery, which for fixed-wing aircraft includes traditional hard-surface landing, soft-surface belly landing, deep-stall auto land, parachute recovery, and net and skyhook recovery.

UAS, regardless of their size, have the same operational components: aircraft, payloads, communications unit, and operator control unit. The aircraft is the platform that flies and carries any required payloads. Payloads are incorporated into or onto the aircraft and are switched to meet particular mission needs. For Department of the Interior missions these payloads will be either a sensor or set of sensors that can acquire the specific type of remotely sensed data that is needed. The aircraft will also carry the payload that is responsible for transmitting live airborne video images, compass headings, and location information to the operator control unit. The communications (or ground) unit, which transfers information between the aircraft and the operator control unit, consists of the hardware and software required to establish both uplink and downlink communications. Finally, the operator control unit both controls and monitors the aircraft and can be operated either by a pilot on the ground or autonomously.

Although development of an increasing variety of UAS platforms substantially increases the usefulness of the technology, it has also flooded the marketplace with options. In one effort to better organize this increasing proliferation of UAS, the Department of Defense created five groupings for UAS based on the technical specifications of platform weight, flight altitude, and airspeed (table 1). There is no similar set of accepted categories on the civilian side, but the Federal Aviation Administration has classified all platforms weighing less than 55 pounds as sUAS.

**Table 1. Unmanned aircraft systems categories as defined by the Department of Defense.**
**[<, less than; >, greater than ; AGL, above ground level; KIAS, knots-indicated air speed; MSL, above mean sea level]**

| Group | Platform weight (pounds) | Flight altitude (feet) | Flight speed | Example platforms |
|---|---|---|---|---|
| | | | | |
| 1 | 0–20 | <1,200 AGL | 100 KIAS | Raven, T-Hawk, Wasp, Draganflyer |
| 2 | 21–55 | <3,500 AGL | <250 KIAS | ScanEagle |
| 3 | <1,320 | <18,000 MSL | <250 knots | Shadow |
| 4 | >1,320 | <18,000 MSL | Any speed | Predator |
| 5 | >1,320 | >18,000 MSL | Any speed | Global Hawk |

Source: Office of the Secretary of Defense, Unmanned Systems Integrated Roadmap 2009–2034: Washington, DC, Government Printing Office, April 6, 2009, p. 96–97.

## Sensor Payloads

Initial sensor payloads transmitted full-motion video using either an electro-optical (EO) camera or a thermal infrared camera. Although this type of remotely sensed data is valuable for many Department of the Interior applications, supporting the full range of potential missions requires a much wider variety of sensors. Fortunately, while unmanned aircraft systems (UAS) technology has been advancing, similar strides have also been made in the miniaturization of sensor packages, resulting in today's wide variety of lightweight and low-cost sensor capabilities including EO, infrared, multispectral, hyperspectral, and synthetic aperture radar.

Sensors detect bands of electromagnetic waves (light) within the electromagnetic spectrum. Passive sensor systems detect reflected electromagnetic waves that already exist in the atmosphere, whereas active systems generate their own source of electromagnetic radiation. This difference means that the use of passive systems can be limited by cloud cover or available daylight, while active sensors can operate in either cloudy conditions or at night.

Optical sensors, which are all passive, use solar illumination or emitted energy and filters, prisms, or diffraction gratings to process spectral bands of electromagnetic waves. EO, thermal infrared, multispectral, and hyperspectral are all types of optical sensors that operate in different ranges of the spectrum. EO sensors (regular cameras) operate in the visible light range within the natural color part of the electromagnetic spectrum. Thermal infrared (heat) sensors detect radiation within the infrared part of the spectrum and can measure temperature differences. Multispectral sensors measure energy in many bands of the electromagnetic spectrum, usually segregated into 3 to 10 different bands. And finally, hyperspectral sensors use either prisms or diffraction gratings for dispersing the measured radiation to collect over 200 narrower and more distinct regions or bandwidths within the electromagnetic spectrum.

Radar is an active sensor system. It generates pulses of electromagnetic radiation and, by using the time it takes for the pulse to bounce back to the antenna, calculates distances and other reflective characteristics. Radar sensors also differ from optical sensors because they operate by looking to the side, not straight down, in order to account for the movement of the antenna between pulses. The physical size of the antenna determines the resolution of radar data; higher resolution is achieved with larger antennas. Unfortunately, this physical dependency has hampered efforts to effectively use radar sensors as payloads in small UAS (sUAS). However, a new form of radar has reenergized miniaturization efforts: synthetic aperture radar uses the movement of the antenna instead of its size to create high resolution imagery.

## UAS Operational Environment—U.S. Department of the Interior

The Department of the Interior (DOI) protects America's natural resources and heritage, honors our cultures and tribal communities, and supplies the energy to power our future. Its breadth of responsibilities includes upholding the Federal government's trust responsibilities to 566 Indian tribes; conserving fish and wildlife and their habitats; managing water supplies for more than 30 million people; and protecting the icons of national heritage (U.S. Department of the Interior, 2011).

Since Congress created the DOI in 1849, it has become the steward for managing more than 500 million acres of surface land, 35,000 miles of coastline, 1.7 billion acres of the Outer Continental Shelf, and a variety of water and underwater resources. These resources supply 23 percent of the Nation's energy and annually host more than 460 million visitors to nearly 5,000 recreational sites (wildlife refuges, parks, monuments, seashore sites, battlefields and other cultural and recreational sites). To support its efforts, the DOI has one of the largest law enforcement workforces in the Federal government (U.S. Department of the Interior, 2011).

The DOI employs approximately 70,000 people at more than 2,400 locations in the following bureaus and offices (U.S. Department of the Interior, 2011):

- Bureau of Land Management
- Bureau of Ocean Energy Management
- Bureau of Reclamation
- Bureau of Safety and Environmental Enforcement
- Indian Affairs
- National Park Service
- Office of Surface Mining, Reclamation and Enforcement
- U.S. Fish and Wildlife Service
- U.S. Geological Survey

## Bureau of Land Management

The Bureau of Land Management (BLM) manages approximately 248 million surface acres and administers and oversees mineral leasing on 700 million subsurface acres of mineral estates underlying other State, private, or federally administered land. These lands account for a substantial portion of U.S. energy production, which includes coal, gas, oil, geothermal power, and installed wind power (U.S. Department of the Interior, 2011).

These diverse and remote public lands—grasslands, forests, high mountains, arctic tundra, and desert landscapes—must be managed for many uses. The BLM manages the use of these public land resources for activities such as energy development, livestock grazing, recreation, and timber harvesting, while also protecting and conserving natural, cultural, and historical resources. The BLM also protects endangered animals, plants, fossils, and archeological, paleontological, and historical sites by sustaining habitat and by documenting stressors on those habitats in order to reduce threats to plant and animal populations.

The BLM also provides wildland fire protection, management, and recovery on vast tracks of public lands in Alaska and the Western States, and it employs uniformed law enforcement rangers to prevent, detect, and investigate crimes affecting resources on public lands.

***How the Bureau of Land Management Uses Remote Sensing Data***

The BLM uses many remote sensing technologies, from satellite-based sensors providing wide-area coverage to simple digital single-lens reflex cameras recording the smallest features on the landscape. It also employs traditional aerial survey techniques that require manned aircraft to be accompanied by a natural resource specialist. Unfortunately, acquiring data by use of manned aircraft can be expensive, and it also adds a personnel safety risk because the areas being flown are often remote or dangerous to access.

To measure erosion of soils, accumulation of sediments in riparian areas, hazardous materials, effects of recreation use, and other developments on BLM lands, random sampling methodology is used on repeated site visits throughout a geographic area. Vegetation changes are irregularly observed during a much longer time span and are mapped on a large scale with Landsat satellites or other remote sensing systems, such as the moderate resolution imaging spectroradiometer, an instrument onboard the National Aeronautics and Space Administration Terra and Aqua satellites (National Aeronautics and Space Administration, 2010). Remote sensing is supplemented through observations along vegetation transects in aircraft or on the ground by using cameras. The challenge to obtaining the increased resolution required for vegetation analysis, either on the ground or in the air, is the vastness of the public lands to be covered and safety concerns with manned aircraft flying at low altitudes.

As part of its wildland fire program, the BLM uses landscape data to map environmental conditions such as types of fuels, locations of fuels, and conditions of fuels. During an active fire real-time data are used determine the status of the fire, weather, and values at risk, in order to monitor fire behavior and measure its size. Typically, aerial observers in aircraft instrumented with forward looking infrared sensors map the fire perimeters and use the imagery to locate hotspots. But manned aircraft missions depend on access to a limited number of aircraft and safe flying conditions, such as little smoke and light winds.

***How the Bureau of Land Management Has Used Data Acquired by Unmanned Aircraft Systems***

The Bureau of Land Management (BLM) is rapidly adopting small unmanned aircraft (sUAS) technology for various natural resource management applications, and it has helped the USGS National Unmanned Aircraft Systems Project Office to develop and test operational processes and procedures. In August 2012 the BLM conducted, in the Big Desert of southeastern Idaho, its first sUAS mission in the National Airspace System. This proof-ofconcept project, requested by the Upper Snake Field Office, tested how well the Raven® and THawk™ sUAS mapped invasive plant species and sagebrush from various altitudes. The goal was to identify the best techniques for identifying an invasive species—in this case, the rush skeletonweed—as a basis for surveying additional areas and obtaining coordinates to use in planning treatments. On the basis of the success of this first mission, an additional mission was flown in August 2013.

In October 2012, the BLM's UAS team conducted flight operations in the San Simon watershed of southeastern Arizona at the request of the Safford Field Office. The Raven was flown to collect images by using high-resolution imaging cameras and high-definition video technology. The six flight days captured more than 26,000 images relevant to rangeland health, infrastructure mapping, engineering site design, erosion headcutting, recreation site management, fence surveys, effectiveness of vegetation treatment, and stream bank modification measurements. This imagery was used to create 3-dimensional models and orthophotographs for each of the project areas.

Other missions in 2013 included mapping a coal-seam fire in Montana, a survey of hazardous material and abandoned mine lands in Arizona, and other collaborative projects such as volumetric calculations of mineral material sites, landfills, and landslides.

The BLM has also successfully partnered with the USGS National Unmanned Aircraft Systems Project Office on many types of small unmanned aircraft missions (see Missions— Greater Sage-Grouse).

## Bureau of Ocean Energy Management

The Bureau of Ocean Energy Management (BOEM) was created on October 1, 2011, as part of the reorganization of the Bureau of Ocean Energy Management, Regulation and Enforcement, formerly the Minerals Management Service. This Bureau "is responsible for managing environmentally and economically responsible development of the nation's offshore resources. Its functions will include offshore leasing, resource evaluation, review and administration of oil and gas exploration and development plans, renewable energy development, National Environmental Policy Act analysis and environmental studies." (Bureau of Ocean Energy Management, Regulation and Enforcement, 2013)

The BOEM is responsible for making sure that all offshore drilling and production activities are performed in an environmentally safe and economically responsible manner. This oversight is provided to all activities in the Outer Continental Shelf, which is a source of considerable oil and gas for the United States. It has approximately 33 million leased acres that account for about 5 percent of the domestic natural gas production and about 21 percent of the oil production (Bureau of Ocean Energy Management, 2013). The BOEM also oversees renewable energy operations on the Outer Continental Shelf, including solar energy, wind energy, wave and ocean current energy, and offshore production of other minerals—such as extracting sand and gravel that is used for coastal restoration projects.

### *How the Bureau of Ocean Energy Management Uses Remote Sensing Data*

The BOEM uses remote sensing data from many sources to monitor energy extraction activities in the Gulf of Mexico. The European Space Agency provides infrared satellite imagery to detect and monitor natural gas flares.

In March 2010, the National Oceanic and Atmospheric Administration began providing maps of oil and other pollutants from a suite of remote sensing satellites, which supports effective monitoring and prompt follow-up. Satellite-based synthetic aperture radar imagery allows rapid assessment of damage to gas and oil infrastructure following hurricanes.

## Bureau of Reclamation

The Bureau of Reclamation manages, develops, and protects water and related resources. Established in 1902 to reclaim arid lands, it has constructed dams, power plants, and canals in the 17 western states to support economic development of the West. Since that time the Bureau has constructed more than 600 dams and reservoirs including Hoover Dam on the Colorado River and Grand Coulee Dam on the Columbia River that provide water, hydroelectric power, and recreation areas (Bureau of Reclamation, 2011a).

The Bureau of Reclamation helps the western states, Native American tribes, and others meet new water needs. It balances a multitude of competing water uses while protecting the environment and the public's investment in water infrastructure. The Bureau is the largest wholesaler of water in the country, operating 348 reservoirs with a total storage capacity of 245 million acre-feet. It serves 31 million people and provides one out of five western farmers (140,000) with irrigation water for 10 million acres of farmland that produce 60 percent of the nation's vegetables and 25 percent of its fruits and nuts. In addition, it is also the second largest producer of hydroelectric power in the western United States. It manages 58 power plants that provide more than 40 billion kilowatt hours annually, enough electricity to serve 3.5 million homes (Bureau of Reclamation, 2011a).

### *How the Bureau of Reclamation Uses Remote Sensing Data*

The Bureau of Reclamation uses a variety of remote sensing technologies such as satellite imagery, aerial photography, and full-motion video supported by field measurement surveys, oblique imagery, global positioning system, and direct observations. Remote sensing instrumentation includes image-based surveillance, optical sensing, and light detection and ranging (lidar).

Remote sensing data are used by the Bureau of Reclamation to map and monitor water delivery infrastructure such as dams, canals, and power plants. "Dams must be operated and maintained in a safe manner, ensured through inspections for safety deficiencies, analyses using current technologies and designs, and corrective actions if needed based on current engineering practices" (Bureau of Reclamation, 2011b). Remote sensing data are also used in geologic investigations that precede design and construction of facilities that deliver water, to detect landslides along creeks, reservoirs and water delivery systems, and to monitor or model river and reservoir water levels, erosion, sediments, water projections, land encroachments, and archeological and cultural historic sites. The Bureau also uses remote sensing technology to study evapotranspiration, water quality, land suitability, flow to habitat, flood, and environmental justice.

Image data also support periodic monitoring of tributary habitats to assess the effectiveness of projects intended to increase beneficial fish habitat. Finally, remote sensing data allow change to be repeatedly monitored in areas too large to cover on foot and too small for traditional aircraft and satellite operations.

#### *How the Bureau of Reclamation Has Used Data Acquired by Unmanned Aircraft Systems*

In 2012 the Bureau partnered with the USGS National Unmanned Aircraft Systems Project Office for small unmanned aircraft missions to monitor downstream river behavior during removal of the Elwha and Glines dams in Olympic National Park, Washington (see Missions—River Restoration).

## Bureau of Safety and Environmental Enforcement

The Bureau of Safety and Environmental Enforcement (BSEE) was created on October 1, 2011, as part of the reorganization of the Bureau of Ocean Energy Management, Regulation and Enforcement, formerly the Minerals Management Service. The BSEE "is responsible for safety and environmental oversight of offshore oil and gas operations, including permitting and inspections, of offshore oil and gas operations. Its functions include the development and enforcement of safety and environmental regulations, permitting offshore exploration, development and production, inspections, offshore regulatory programs, oil spill response and newly formed training and environmental compliance programs" (Bureau of Ocean Energy Management, Regulation and Enforcement, 2013).

#### *How the Bureau of Safety and Environmental Enforcement Uses Remote Sensing Data*

The BSEE responds to oil spills, which commonly develop in isolated offshore locations. In these situations, remote sensing data are irreplaceable for monitoring purposes. The National Commission's report of the Deepwater Horizon disaster said that drilling can continue in the Gulf of Mexico with improved oversight, but it questioned whether anyone would be capable of dealing with a similar accident off the coast of Alaska (National Commission on the BP Deepwater Horizon Oil Spill and Offshore Drilling, 2011). "The

Alaskan Arctic is characterized by extreme cold, extended seasons of darkness, hurricane-strength storms, and pervasive fog— all affecting access and working conditions" (National Commission on the BP Deepwater Horizon Oil Spill and Offshore Drilling, 2011). The commission further stated that the Federal government should launch an extensive effort to boost oil spill response capabilities in the area.

Addressing these concerns and continuing pressure to pursue energy independence will push government, including the BSEE, to leverage new, innovative, and more efficient remote sensing technologies for responding to disasters like oil spills.

## Indian Affairs

Indian Affairs, which contains the Bureau of Indian Affairs (BIA) and the Bureau of Indian Education, is the oldest bureau of the Department of the Interior. Established in 1824, it provides services to approximately 1.9 million Native Americans who belong to 566 federally recognized American Indian tribes and Alaska Natives. The BIA is responsible for fulfilling Indian trust responsibilities, to "enhance the quality of life, to promote economic opportunity, and to carry out the responsibility to protect and improve the trust assets of American Indians, Indian tribes, and Alaska Natives" (U.S. Department of the Interior Indian Affairs, 2011). The mission of the Bureau of Indian Education is to "... provide quality education opportunities from early childhood through life in accordance with the tribes' needs to cultural and economic well-being in keeping with the wide diversity of Indian tribes and Alaska Native villages as distinct cultural and governmental entities" (U.S. Department of the Interior Indian Affairs, 2011).

As part of its responsibilities, the BIA manages natural resources on 55 million surface acres and 57 million acres of subsurface minerals estates held in trust for American Indians, Indian tribes, and Alaska Natives. It also oversees many other programs, such as economic development programs in some of the most isolated and economically depressed areas of the United States, law enforcement and detention services, administration of tribal courts, implementation of land and water claim settlements, housing improvement, disaster relief, replacement and repair of schools, repair and maintenance of roads and bridges, and the repair of structural deficiencies on high hazard dams (U.S. Department of the Interior Indian Affairs, 2011).

***How the Bureau of Indian Affairs Uses Remote Sensing Data***

The Bureau of Indian Affairs (BIA) applies remote sensing to activities such as land use planning, responding to nonpoint-source pollution affecting subsistence hunting and fishing, effects of climate change such as sea level rise for coastal tribes, location and identification of potential dam hazards, and the generation of digital terrain data for the use of open-channel hydraulics. In 2011, the BIA started integrating remote sensing into forest inventory and management planning in remote, low timber value, and inaccessible areas.

## National Park Service

The mission of the National Park Service is to "preserve unimpaired the natural and cultural resources and values of the National Park System for the enjoyment, education, and inspiration of this and future generations and to cooperate with partners to extend the benefits of resource conservation and outdoor recreation throughout this country and the world" (National Park Service, 2010a). The Park Service oversees 393 national parks that comprise 84 million acres of land, 4.5 million acres of water (oceans, lakes and reservoirs), more than 85 thousand miles of rivers and streams, and 43 thousand miles of shoreline. These areas contain more than 68 thousand archeological sites, 27 thousand historic structures, more than 121 million objects in their museums, and more than 2,400 historic and nearly 600 natural national landmarks (National Park Service, 2010b).

In national parks, the Park Service works to minimize the loss or degradation of culturally significant material, monitor air quality, manage water and aquatic resource, and manage fires. Park rangers enforce laws that protect people and the parks and provide emergency services such as fire response and search and rescue. The U.S. Park Police prevent and detect criminal activity such as poaching, dumping, and encroachments, and protect U.S. artifacts. They apprehend individuals in national parks suspected of committing offenses against Federal, State and local laws, provide personal protection to the President of the United States and visiting dignitaries, and protect monuments, memorials and museums in the national parks.

***How the National Park Service Uses Remote Sensing Data***

The National Park Service collects data in the continental United States and Alaska through satellites, fixed-wing aircraft and rotorcraft, and human observations. Associated with these methods are various sensors (for example, optical and thermal) that are either fixed to the chosen platform or hand held. During wildland or prescribed fires, fire fighters need information such as fire location, perimeters, active edges, direction of spread, burn intensity, and smoke plumes and dispersal. Already-burned areas are examined (Is the vegetation burned, scorched, or unburned?) as is the active fire line (Where is the flaming front? What is the residual heat?). The location of homes and other structures or infrastructure at risk guide decisions about distribution of firefighting resources.

Owing to the expanse of the national park system, incidents calling for intervention by park police or park emergency services typically are spread among many jurisdictions, and they may involve two or more parks, States, wilderness areas, and rural and urban settings. For instance, invasive plant species and marijuana farms can be located and monitored by satellite and by on-ground sampling of accessible areas, such as road rights-of-way and streams. But, if surveys are conducted only in areas that are accessible, the accuracy of vegetation inventories is limited because they are not random and may not extrapolate to the rest of the landscape. Remote sensing offers a safe alternative for scouting invasive or illegal crops, especially in rugged terrain.

#### *How the National Park Service Has Used Data Acquired by Unmanned Aircraft Systems*

The National Park Service has successfully partnered with the USGS National Unmanned Aircraft Systems Project Office on many types of small unmanned aircraft missions (see Missions—Abandoned Solid Waste, Resource and Vegetative Species, and River Restoration).

## Office of Surface Mining Reclamation and Enforcement

The Office of Surface Mining Reclamation and Enforcement (OSM) implements two programs resulting from the Surface Mining Control and Reclamation Act of 1977: (1) the Abandoned Mine Lands program to reclaim lands containing mines abandoned before 1977, and (2) regulation of current mine operations by developing standards and procedures for approving mining (Office of Surface Mining Reclamation and Enforcement, 2008; Office of Surface Mining Reclamation and Enforcement, 2010).

The Federal government owns large acreages of land and coal reserves, primarily in the western United States. The OSM oversees States and tribes that are the primary regulators of mining operations, and it has direct responsibility for mine lands in the remaining States, reservations, and Federal lands. Reclaiming lands that contain abandoned mines is challenging. Abandoned mines may still pollute streams with effluent that contains iron, manganese, aluminum, and other metals derived from bedrock minerals rich in iron sulfide, such as pyrite. Iron hydroxide or sulfuric acid in the effluent is a particular concern. These mines can also harbor open mine shafts and deadly mine gases and may be surrounded by steep, unstable slopes or ground prone to collapse over aged underground tunnels.

### *How the Office of Surface Mining Uses Remote Sensing Data*

The OSM derives most of the data it needs from on-the-ground inspections augmented by satellite imagery or aerial photographs from manned aircraft. However, each mode has its drawbacks. Ground observation of all of the large number of mines is almost impossible. Some are in remote or potentially dangerous locations, and limited access or inclement weather may also interfere. Abandoned mines are difficult to detect from satellite images, and aerial surveys from small aircraft and helicopters are expensive.

The data derived from mine site inspections are used to address topics such as water quality, hazardous conditions, terrain, wildlife habitats, postmining land use, and cultural features. Active mines are inspected before mining starts; during mining they are partially inspected bimonthly and fully inspected quarterly. Inspections of abandoned mine land should be ongoing and continuous; however, because existing abandoned mine land inventories are not complete and may be inaccurate, the priority is simply determining the location of these mines.

### *How the Office of Surface Mining Has Used Data Acquired by Unmanned Aircraft Systems*

The OSM has successfully partnered with the USGS National Unmanned Aircraft Systems Project Office on many types of small unmanned aircraft missions to inspect abandoned mine lands and survey surface mine sediment ditches (see Missions—Contour Surface Mine and Abandoned Mine Lands).

## U.S. Fish and Wildlife Service

The U.S. Fish and Wildlife Service (USFWS) oversees the conservation, protection, and enhancement of fish, wildlife and plants, and their habitats. This oversight applies to the 150 million acres in the National Wildlife Refuge

System, which is devoted to the protection and conservation of fish and wildlife and their habitats. More than 47 million people visit one of the 561 refuges each year to hunt, fish, observe wildlife, and participate in other outdoor recreation activities (U.S. Department of the Interior, 2011).

Protecting endangered and threatened species, overseeing their recovery, and protecting habitats is also a priority for the USFWS. It assists in the restoration of habitats harmed by pollution, evaluates water quality, responds to oil and chemical spills, evaluates pesticide effects on fish and wildlife during pesticide registrations, and investigates the effects of contaminants. It is also responsible for implementing and enforcing environmental laws, such as the Endangered Species Act, Migratory Bird Treaty Act, Marine Mammal Protection Act, North American Wetlands Conservation Act, and Lacey Act.

To contribute to the health and sustainability of migratory birds, the USFWS monitors, assesses, and annually surveys both game and nongame birds. However, the complex pathways of migratory birds, represented by their flyway systems, require cooperation of regional, national, and international scope (Birds and Nature, 1998). The USFWS also participates in international treaty negotiations and coordinates efforts to reduce bird mortalities resulting from collisions with equipment and structures, by-catch, pesticide, and other human-related causes.

***How the U.S. Fish and Wildlife Service Uses Remote Sensing Data***

The USFWS is charged with protecting the status of more than 800 species of migratory birds that spend at least some of their lives in the United States (U.S. Fish and Wildlife Service, 2004). Because migratory birds are ubiquitous, accurate estimates require nearly simultaneous surveys (to avoid double counting) throughout broad swaths of their range. Information may be gathered by ground counts along transects, estimates based on visual observations of bird abundance from low-altitude manned aircraft, or aerial videography and satellite imagery. It is unlikely that ground resources will ever be sufficient to monitor all required species, which makes aerial methods the only feasible means of obtaining the required data. However, manned aerial surveys present safety concerns. Even in flat terrain, aerial surveys are flown at low enough altitudes that they can conflict with obstacles on the ground such as wind turbines and communication towers.

The USFWS also restores wetland habitat by improving water quality and by controlling the spread of exotic plants. Water-bird colonies are typically monitored by aerial photographs that are coupled with other geospatial data to estimate bird counts. Digital aerial photographs are also used to assess wetland habitat and monitor easements; they require an on-the-ground spatial resolution of 1.3 meters per pixel. Monitoring easement-protected wetlands can involve tens of thousands of wetlands scattered in several states. These photos are taken by small manned aircraft at heights of 1,800 to 9,500 feet above mean sea level and then combined with other data to enable measurements of interest.

***How the U.S. Fish and Wildlife Service Has Used Data Acquired by Unmanned Aircraft Systems***

The U.S. Fish and Wildlife Service has worked with the USGS National Unmanned Aircraft Systems Project Office on several small unmanned aircraft missions (see Missions— Sandhill Crane, and Waterfowl Surveys).

## U.S. Geological Survey

The U.S. Geological Survey (USGS) provides impartial scientific information on the health of our ecosystems and environment, the natural hazards that threaten us, the natural resources we rely on, the effects of climate and land-use change, and the core science systems that help us provide timely and relevant information (U.S. Department of the Interior, 2011). The USGS serves the Nation by studying our environment on a national scale and by long-term monitoring and assessment of natural resources.

Because it has no regulatory or management mandate, the USGS can provide impartial science that provides the scientific foundation upon which policymakers, natural resource managers, and the public make informed decisions about the management of natural resources. This scientific foundation also provides rich data, such as the largest archive of remotely sensed land data in the world, the geospatial baseline of the U.S. (topography, natural landscape, and built environment), and geologic maps and 3-dimensional geologic frameworks.

***How the U.S. Geological Survey Uses Remote Sensing Data***

The USGS uses a variety of remote sensing techniques to observe changes to the landscape. Aerial photography is generally used for research on submerged aquatic vegetation, because it is an effective way to find these patches of vegetation while not disturbing the marshes. Manned aircraft surveys are also used to estimate the population of invasive species. However, manned aircraft is expensive, constrained with respect to time, dangerous owing to the requirement to fly low and slow, and disruptive to wildlife. Additionally, the scale of a study may be so small (a single pond or marsh) as to make traditional aircraft or satellite operations ineffective.

Land surface deformation caused by earthquakes, landslides, debris flows, land subsidence, and glacier motion are measured by calculating point-position land changes with geodetic techniques (such as ground-based tripod, lidar, global positioning system, leveling, airborne lidar, and multispectral airborne) and satellite interferometric synthetic aperture radar imagery. Once the imagery is collected, land surface change is measured by detecting changes in related images such as the offset of pixels in raster images. Measurements are taken at various scales during time periods ranging from minutes to years.

Emergencies caused by earthquakes, volcanic eruptions, and floods require real-time information throughout territory that can be vast, rough, and remote. Even with use of several geodetic and imaging techniques, it is not always possible to provide rapid site inspection during disasters or feasible to use ground-based observers. To monitor volcanoes in the United States and its territories, the USGS collects and interprets real-time and near real-time geophysical data that are integrated with satellite imagery collected by other organizations, such as National Aeronautics and Space Administration and the National Oceanographic and Atmospheric Administration.

Obtaining cost-effective high-resolution imagery of a large area remains a challenge. Satellite imagery may have too-low resolution, and ground-based imagery is inefficient for large areas. One high-resolution image for a large area can be costly, and the cost of repeated images needed to detect changes in the landscape can quickly become prohibitive.

#### *How the U.S. Geological Survey Has Used Data Acquired by Unmanned Aircraft Systems*

In the spring of 2008, the USGS established the National Unmanned Aircraft Systems Project Office (NUPO) in Denver, Colorado, to support the integration of unmanned aircraft systems (UAS) technology into the Department of the Interior. Since its inception, the NUPO has successfully completed several small UAS missions (see Missions).

## Office of the Secretary of the Interior

The Office of the Secretary of the Interior consists of the Departmental Offices for the Secretary; Deputy Secretary; Solicitor; Inspector General;

Assistant Secretary for Policy, Management and Budget; and the Assistant Secretary for Insular Affairs. The Deputy Assistant Secretary for Public Safety, Resource Protection and Emergency Services also resides in this office and oversees the Office of Aviation Services (OAS) (U.S. Department of the Interior, 2011). The OAS, which was established by the Secretary of the Interior on July 1, 1973, manages functions related to aircraft services and facilities, particularly their safety. The OAS also works to increase the efficient and economical operation of aircraft, including UAS, for the Department of the Interior (DOI) (Office of Aviation Services, 2013a). It was recognized as the best aviation program in the Federal government (2008), has achieved Gold Standard certification for Federal aviation programs continuously since 2008, and is the only Federal aviation services office to have achieved the ISO 9001–2008 international quality certification.

The OAS also partners with the USGS National Unmanned Aircraft Systems Project Office (NUPO). Whereas the NUPO supports mission requirements and researches sensors, the OAS focuses on safe unmanned aircraft operations. In addition, the two offices determine how to best use unmanned aircraft systems (UAS) services: procuring platforms, establishing commercial contracts, coordinating with other governmental entities, and integrating UAS into the National Airspace System.

Summarized in Table 2 are all earth observations made by DOI bureaus in which remote sensing plays a role.

**Table 2. Types of remote sensing activities performed by the Department of the Interior**

**[BIA, Bureau of Indian Affairs; BLM, Bureau of Land Management; BOEM, Bureau of Ocean Energy Management; BSEE, Bureau of Safety and Environmental Enforcement; BOR, Bureau of Reclamation; NPS, National Park Service; OSM, Office of Surface Mining, Reclamation and Enforcement; USFWS, U.S. Fish and Wildlife Service; USGS, U.S. Geological Survey]**

| Observations | | | | | Bureau | | | | |
|---|---|---|---|---|---|---|---|---|---|
| | BIA | BLM | BOEM | BSEE | BOR | NPS | OSM | USFWS | USGS |
| Archaeological inventory | • | • | | | • | • | • | • | • |
| Dam inspection | • | • | | | • | • | • | • | |

**Table 2. (Continued)**

| Observations | | | | | Bureau | | | | |
|---|---|---|---|---|---|---|---|---|---|
| | BIA | BLM | BOEM | BSEE | BOR | NPS | OSM | USFWS | USGS |
| Geophysical surveys | | • | • | • | • | | • | | • |
| Global change | • | • | • | • | • | • | • | • | • |
| Hydrology | • | • | • | • | • | • | • | • | • |
| Disturbed-surface monitoring | • | • | | | • | • | • | • | • |
| Law enforcement | • | • | | | • | • | | • | |
| Mine inspection | • | • | | | | • | • | | |
| Natural resources | • | • | • | • | • | • | • | • | • |
| Road inspection | • | • | | | • | • | • | • | |
| Vegetation and habitat | • | • | • | • | • | • | • | • | • |
| Volcano hazards | | | | | | | | | • |
| Wildland fire | • | • | | | • | • | | • | |
| Wildlife management | • | • | • | • | • | • | • | • | • |

## Unmanned Aircraft Systems Operations

The Department of the Interior (DOI) has begun to set up a cost-efficient unmanned aircraft systems (UAS) operation. Early on, the Office of Aviation Services and the newly created USGS National Unmanned Aircraft Systems Project Office evaluated UAS and developed an operational strategy. They partnered with the U.S. Army to acquire access to its fleet of AeroVironment Raven RQ-11A small UAS (sUAS), then established training requirements and provided classes that resulted in a cadre of DOI personnel certified as Raven operators. These trained pilots were able to use this fleet for proof-of-concept missions that supported the development and testing of approval procedures, operational processes, and cost:benefit information.

A comprehensive UAS strategy is tailored to the various mission, funding, personnel, and infrastructure levels of the DOI and includes the following:

1) Focus on sUAS, which align better with DOI's decentralized missions and with the Department's funding and infrastructure levels.

2) Leverage available excess Department of Defense (DOD) sUAS to minimize procurement, training, and support costs.
3) Establish partnerships with Federal agencies that may be able to support DOT missions in need of larger UAS platforms.
4) Test and evaluate various sUAS technologies in order to develop long-term UAS requirements and strategy.
5) Based on the requirements and strategy developed in the above step, buy or contract for UAS capabilities that cannot be met either through excess DOD sUAS or those available through partnerships with other Federal agencies.

## Organization

For more than 50 years, the Department of the Tnterior (DOT) has used manned aircraft to support its missions, so using unmanned aircraft systems (UAS) technology is simply a logical next step. The evolution of UAS organizational support within DOT started with the creation of the USGS National Unmanned Aircraft Systems Project Office (NUPO) and involvement of the Office of Aviation Services (OAS). Working together, these two groups arranged access to the Raven small UAS (sUAS) surplused by the U.S. Army and the establishment of a Federal Aviation Administration (FAA) approved Raven operator training course. The OAS then provided initial training classes that resulted in more than 50 DOT scientists becoming certified Raven operators.

As with manned aircraft, UAS activities are regulated by OAS: operator training, certification of platform airworthiness, fleet management, acquisitions, and safe operation. An OAS memorandum outlines several requirements for use of UAS by the DOT (Office of Aviation Services, 2013b), and the OAS currently manages the Ravens and Honeywell's T-Hawk Micro Air Vehicle sUAS that have been provided by the U.S. Army.

The USGS NUPO, one of the first UAS organizations within DOT, investigated the potential for UAS to support specific remote sensing needs. As part of successful proof-ofconcept missions, the NUPO developed processes for receiving FAA Certificate of Authorization for missions, and all of the underlying permissions such as radio spectrum approval. During these tasks NUPO worked with other bureaus to establish additional operational UAS groups and increase the overall use of the technology. Both the OAS and the USGS NUPO continue to work towards full integration of UAS into the DOT.

## Operator Training

The Federal Aviation Administration (FAA) allows the Department of the Tnterior (DOT), as a public agency, to self-certify operators of its unmanned aircraft systems (UAS). Tnitially the Office of Aviation Services (OAS) developed, for FAA review, training materials for both the Raven and T-Hawk. After the FAA determined that these materials met all required regulations, the OAS was allowed to train and certify DOT personnel to operate either the Raven or T-Hawk small UAS (sUAS). FAA requirements allow only certified personnel with a current class 2 medical certificate to operate sUAS in the National Airspace System. The FAA may also require additional items, such as a private pilot license, determined by the nature of the sUAS mission and established through the FAA's Certificate of Authorization.

## Planning Missions by Using a Business-Case Model

Addressing the cost of unmanned aircraft systems (UAS) compared with their benefit is part of their implementation. A business-case model is used to help determine whether a proposed UAS mission would cost effectively provide the required remote sensing data. First, two elements are determined: the remote sensing requirement (the type of mission and the data needed), and UAS options. Then, it should be possible to determine if a UAS mission can provide the needed data and, if so, whether the cost of using UAS is reasonable as compared with alternatives. At this point, whether to proceed with a UAS mission can be decided.

The business case first states the final objective of the mission (for example, establishing bird populations) and then determines the type of remotely sensed data needed (for example, image data from a Raven flight can be used to identify bird populations; see fig. 2). Bureaus within the Department of the Interior each have many monitoring objectives to meet. Some of these objectives are mandated by statute, but others attempt to quantify aspects of complex Earth systems and their underlying scientific variables. The U.S. Fish and Wildlife Service may meet its mandated monitoring objectives by surveying migratory birds and by counting nests and the number of eggs in a nest, but it might also use this same information to help establish leaching of pesticides into groundwater and their long-term effect on a region's ecology. Such "layered" mission needs require that all objectives be identified at the outset, after which establishing the remote sensing data requirements should be relatively easy. For instance, if the only objective is to monitor the population of migratory birds, then the only data needed are current imagery that can concurrently cover the roost area at a scale that allows individual birds to be counted.

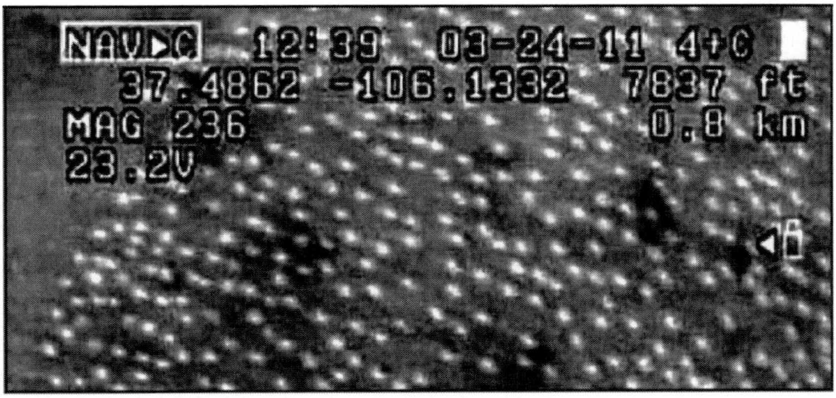


Figure 2. Image from a Raven flight showing individual birds (white dots) which makes it useful for calculating wildlife populations.

After all data requirements are established, the next step is to determine what data or technology are available to meet these needs and what are their associated costs. This can be done by addressing these questions: Are there existing data sets? Are there current manned platforms or satellite data sets that could provide the necessary data? Are these data an emerging or previously unsatisfied requirement? For satellites and manned aircraft, answers to these questions will be based on years of established operational information. However, because UAS technology is relatively new, operational information will be limited to results from ongoing proof-of-concept missions. To date (2014) the National Unmanned Aircraft Systems Project Office has access to Raven and T-Hawk small UAS (sUAS), which can acquire data from multispectral imagers and from thermal infrared and natural color electro-optical sensors.

Agreements with the Department of Homeland Security Customs and Border Protection have allowed access to their UAS Predator® fleet, which can acquire data from electro-optical, thermal infrared and synthetic aperture radar sensors. Information from completed proof-ofconcept missions (documented in this Roadmap), can be used to identify the types of data that can be acquired with this technology and the associated operational costs. Although proof-of-concept missions continue, enough information is now compiled to establish that UAS technology can effectively gather many types of required data, and the cost of doing so. Therefore, if traditional methods and UAS options both exist for gathering data, the costs should be compared. If no traditional methods exist, then whether costs of using UAS are within the mission budget should be the determining factor.

## Privacy

The Department of the Interior (DOI) has decades of experience in the proper collection, use, control, and retention of aerial data from the many manned aircraft that it employs. It uses the same data collection, use, control, and retention policies for its unmanned aircraft systems (UAS) flights as it does for its manned aircraft missions.

Although UAS have proven their effectiveness, concerns about the potential of UAS to infringe on an individual's right to privacy threaten to overshadow its benefits. UAS used by DOI support activities such as scientific research, monitoring environmental conditions, analyzing the effects of climate change, responding to natural hazards, understanding rates and consequences of landscape change, and related land management. The potential benefits are irrefutable. However, privacy concerns must be dealt with effectively if a DOI bureau expects the public to continue its support of UAS, especially as the potential for additional routine uses could increase privacy concerns.

All UAS missions are now and will be in the future in full compliance with Federal laws and DOI policies and procedures. Images collected with UAS sensors are handled and retained according to industry standards, consistent with images collected with any of the DOI's remote sensing systems. The UAS missions are subject to professional standards, codes of conduct, case law, and with the public's trust in mind.

The DOI has also taken the following actions related to an individual's reasonable expectation of privacy:

- UAS operate primarily over public lands.
- All operations must be in support of mandated missions of the DOI ("proper use").
- Written permission must be obtained from landowners if UAS missions target specific activity over their lands.
- Standard FAA-approved procedures are followed for aircraft operations (such as special use permits, range approval letters, and safety management systems).
- Open and transparent UAS operations: UAS mission schedules are listed on the USGS National Unmanned Aircraft Systems Project Office website [http://uas.usgs.gov], and local press and landowners are invited to observe the operations, if appropriate.
- Any privacy concerns will be addressed in consultation with the Office of the Solicitor.
- Unauthorized use of UAS will result in severe penalties.

## Federal Aviation Administration Certificate of Authorization

To fly any public unmanned aircraft in the National Airspace System (NAS), one must first obtain a Certificate of Authorization (COA) from the Federal Aviation Administration (FAA). The COA is an approval issued by the FAA to a public operator granting permission to fly an unmanned aircraft mission over a specific area for a defined period of time. Under current FAA policy, only U.S. government (Federal, State, or local) entities and universities can apply for a COA.

First, supporting documents such as the operational plan, risk management plan, platform airworthiness certification, pilot qualifications, radio spectrum approval, and communication plan must be completed. Next, a COA request is entered into the FAA's online, password-protected COA system. Access to this system is coordinated by the bureau aviation manager, so any potential operators will need to request access prior to submitting their first COA request. Once access has been provided the operator will be required to fill in all portions of the request form, as well as attach required supporting documents:

- Mission description
- Mission flight operations
- Class of airspace to be flown

- Date of operation
- Radio frequency spectrum analysis approval
- Flight operations plan including maps and authorized area extents
- Special circumstances (special use permits)
- UAS system descriptions
- Performance characteristics (launch and recovery)
- Platform Airworthiness Certification
- Operating procedures (emergency, lost link, lost communications)
- Flight aircrew qualifications (medicals, training, currency, and proficiency)

Once the complete package has been entered, the operator will notify the bureau aviation manager, who will confirm that all required information is present. Then the bureau aviation manager will ask the Office of Aviation Services (OAS) UAS specialist to submit the request to the FAA. Finally, the FAA reviews all operational and technical information in the COA package.

According to current (2014) procedures, the FAA has 60 business days after COA submittal to complete three review phases: administrative, safety and coordination, and activation or release of the COA. During the administrative phase, the FAA ensures that all required documents are in place and overall project feasibility. During the safety and coordination phase, the FAA reviews aircraft, flight crew, and operational procedures for compliance with current FAA National Policy N 8900.227 Unmanned Aircraft Systems (UAS) Operational Approval (Federal Aviation Administration, 2013). In the third phase, the FAA approves or denies the COA. During its review, the FAA may request additional provisions or limitations in order to ensure that UAS will operate safely with other airspace users. These additional stipulations could limit the altitude or time of day of a flight or require more observers or the presence of a certified manned aircraft pilot. If the FAA requests any changes to the COA (either additional provisions or edits) or denies the request, then the COA is released back to the submitting operator. Otherwise the FAA approves the COA and allows the requesting operator to proceed with the UAS activity as specified in the request.

An approved COA can also be renewed or extended on a case-by-case basis, or allowed to expire when no additional flights are required. A COA can be issued for a maximum of 24 months, or until the earliest expiration date of any of the supporting documentation (for example, airworthiness, radio spectrum approval, special use permits). The possibility of having overlapping approvals, such as an approved COA that extends past the approved spectrum-use date, makes it essential that those using a COA constantly monitor not only the status of each approved COA but also all of the underlying documents.

The recent establishment of a memorandum of agreement between the Department of the Interior and the FAA altered the above process when a small UAS mission is being flown in Class G airspace (appendix 4). This memorandum allows the DOI to fly sUAS in Class G airspace after providing necessary flight information and notifying the FAA before the flight. The memorandum focuses on routine low-altitude operation (less than 400 feet above ground level) of sUAS. Night flights are also allowed since aircraft position lights allow easier tracking as compared with tracking under daytime visual flight rule conditions, and night flights also benefit from reduced aviation traffic. Similar memorandums of understanding have been established between the FAA and other Federal agencies such as the Department of Justice and NASA.

## Radio Spectrum Approval

One of the underlying requirements of a COA is approval to use the radio frequencies needed during the mission. The electromagnetic spectrum is a crowded neighborhood, so any unmanned aircraft systems (UAS) communication must maintain appropriate separation from adjacent or interfering signals. This management and coordination activity, chartered by the Communications Act of 1934 (Public Law 73-416, 48 Stat. 1064) and later revised by the Telecommunications Act of 1996 (Public Law 104-104, 110 Stat. 56), has become increasingly complex as demand for spectrum has grown exponentially. The Office of Spectrum Management manages the Federal government's use of the radio frequency spectrum.

The Department of the Interior (DOI) and its bureaus coordinate all spectrum management through the National Spectrum Management Office of the Office of Chief Information Officer of the Office of Secretary of the Interior. It is within this specialized office that all "licensing" for DOI activities is coordinated with the National Telecommunications and Information Administration (NTIA). Just as the Federal Communications Commission manages licensing for nongovernment and commercial interests, the NTIA allocates spectrum authorizations to Federal government agencies such as the DOI.

Operational use of the Ravens, and any other small UAS acquired from the military that operate on the military spectrum, also requires coordination with the military spectrum group. However, this additional coordination will not be required for newer nonmilitary systems that use either public, dedicated Federal (excluding the Department of Defense), or DOI-only frequencies.

## Systems

### *Raven—Fixed-Wing Small Unmanned Aircraft*

The Department of the Interior (DOI) obtained access to a fleet of Raven RQ-11A small UAS (sUAS) (fig. 3) through the efforts of the National Unmanned Aircraft Systems Project Office (NUPO), which worked with the Office of Aviation Services (OAS) to establish a strategic partnership with the U.S. Army. The hand-launched Raven system, produced by AeroVironment, is used primarily as a reconnaissance and surveillance tool. It is battery powered, lightweight at 4 pounds, and relatively quiet, which makes it ideal for use in the rough country or wildlife refuges that the DOI monitors.

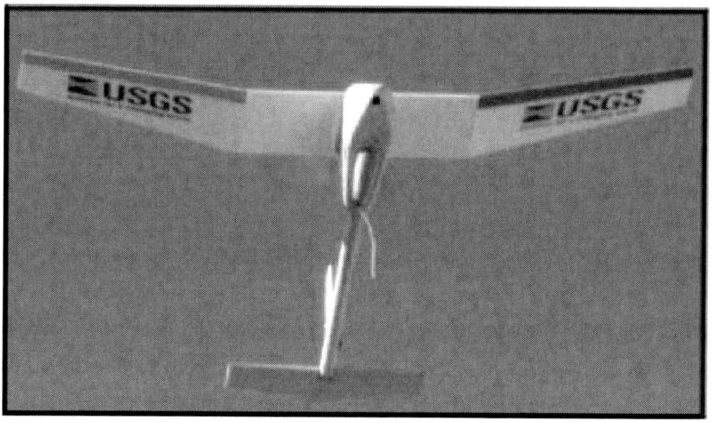


Figure 3. Photograph of a AeroVironment RQ-11A Raven fixed-wing small unmanned aircraft.

During flight, this system transmits analog full-motion video and associated geospatial metadata such as compass headings and global positioning system location information to a ground control unit and a remote video terminal. A trained operator uses the control unit to monitor and control the Raven's flight. The real-time streamed video and metadata are stored during flight as digital files of base imagery that can be used for analyses after the mission. The original sensor packages provided with the Ravens includes side- and forward-looking electro-optical color, side-looking thermal infrared and forward-looking thermal infrared sensors. In addition the Raven has been modified, in coordination with the OAS, so that it supports newer digital sensors, such as GoPro Hero high definition cameras.

### *T-Hawk—Vertical Take-Off and Landing Small Unmanned Aircraft*

After successful use of the Ravens, the U.S. Army also provided the Department of the Interior with access to the RQ-16A T-Hawk micro air vehicle (fig. 4). This small UAS was produced by Honeywell as a reconnaissance and surveillance system and has hover, persistent stare, and vertical launch-and-land capabilities. It streams digital full-motion video, in either electro-optical or infrared spectrum, and it has been modified to support additional high-definition sensors. It is heavier than the Raven at 18 pounds, operates on gasoline instead of batteries, and is very loud (often called a flying lawn mower). However, its ability for vertical takeoff and persistent stare make it an ideal platform for site inspections and damage assessments.

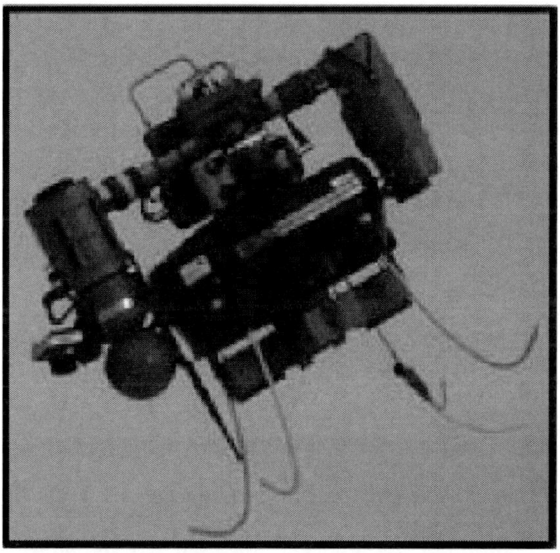

Figure 4. Photograph of a Honeywell RQ-16A T-Hawk vertical takeoff and landing small unmanned aircraft.

***Predator—Large Unmanned Aircraft***

The Department of the Interior (DOI) has determined that large unmanned aircraft systems (UAS) are too costly to own and operate. However, the data generated by these systems, which can use more advanced sensors over a larger area, have obvious benefits. Therefore the DOI has partnered with Federal departments that already possess UAS beyond those of the DOI. The Department has already worked with the Customs and Border Protection (CBP) to acquire access to remotely sensed data acquired during CBP Predator missions.

CBP operates a fleet of Predator B aircraft that were manufactured by General Atomics Aeronautical Systems as long-endurance high-altitude UAS. The Predator, which can operate for 20 hours at an altitude of 20,000 feet, carries state-of the-art sensors including electrooptical, thermal infrared and synthetic aperture radar. Because of its unique combination of operational capabilities, payload capacity, mission flexibility, potential to accommodate new sensor packages, and its safety and performance record, this system is ideal for use as a multimission system by various Federal customers. In an effort to establish additional Federal customers for the imagery provided by the Predator, the CBP approached USGS and offered to image areas of interest to DOI on a noninterference basis or at USGS's request (for example, to obtain images of a natural disaster). These data can be viewed as they are being gathered, and potential data users can communicate directly with the operators to ask for specific data captures while the Predator is in flight. This system delivers data for analysis in minutes, which is a substantial improvement over more traditional data-dissemination capabilities.

## Sensor Data

The raw images captured by the sensors on unmanned aircraft systems (UAS), including full-motion video and still-frame photos, can produce many types of geospatial data for scientists to use in their studies and analyses (fig. 5).

Because of the volume of raw data, both the data storage and processing power needed to produce the geospatial products can be substantial. For example, current cameras capture still images with a 12-megapixel sensor that produces a compressed .jpg output file of approximately 1.5 megabytes (Mb). The full-motion video is typically shot at 1,080 pixels with 30 frames per second, which translates to roughly 100 Mb per minute of compressed MP4 video, and although flight durations can vary, most will exceed 30 minutes. As the technology of small, lightweight sensors advances, larger "small" sensors will be available that can produce as much as or even more than 15 Mb of raw, uncompressed data.

Currently GoPro Hero 2 and 3, which are 1,080 pixel high-definition cameras, are used to capture still-frame and video imagery. The Canon SX260HS, a global positioning system (GPS) enabled red-green-blue camera, and the Canon SX230HS, a GPS enabled red-green-blue and color infrared camera, are also used to capture imagery. These cameras, which are very small and lightweight, can be easily attached to either the Raven or T-Hawk and alternated as needed. GoPro cameras attached to Ravens obtain GPS data only by means of the onboard Raven telemetry-collecting hardware, which makes it difficult to tie GPS data to the GoPro still images owing to inconsistencies in the timestamps. In order to overcome this problem, physical ground control points are scattered around the mission site to be used during image postprocessing. Canon cameras that are being used for T-Hawk missions have built-in GPS sensors that allow an average spatial accuracy of 8–12 meters. This feature allows faster and more accurate processing of raw image data. It is believed that combining onboard camera GPS and ground control could increase this accuracy to centimeters.

Raw image data, once acquired, must be orthorectified to map standards before it can be used with image data from other platforms or to produce geospatial image products. This process uses GPS data acquired during the flight to provide precise geolocation information that calibrates the imagery. Calibrated data can subsequently be orthorectified to a digital surface defined by digital terrain models or used in any number of geospatial analyses (fig. 6). Although each sensor can have unique requirements for preprocessing, in most cases the output signals of the sensor must be radiometrically corrected to physical units. Calibration of optical sensors uses known targets with ground measurements of reflectance properties, and the outputs are radiances by spectral band or wavelength. Atmospheric correction models can correct these radiances to reflectance values and compare them with reflectance observations from other sources. Optical sensors may have deficiencies such as variation in detector-to-detector calibration, dropouts, systematic noise patterns, and nonsymmetrical optical transfer functions. A high-quality camera calibration that quantifies the distortions of the camera lens system has been a fundamental requirement for precise and accurate extraction of topography (for example, terrain and 3-dimensional surfaces) or planimetric features (for example, road centerlines, streams, or vegetation boundaries) from stereoscopic imagery.

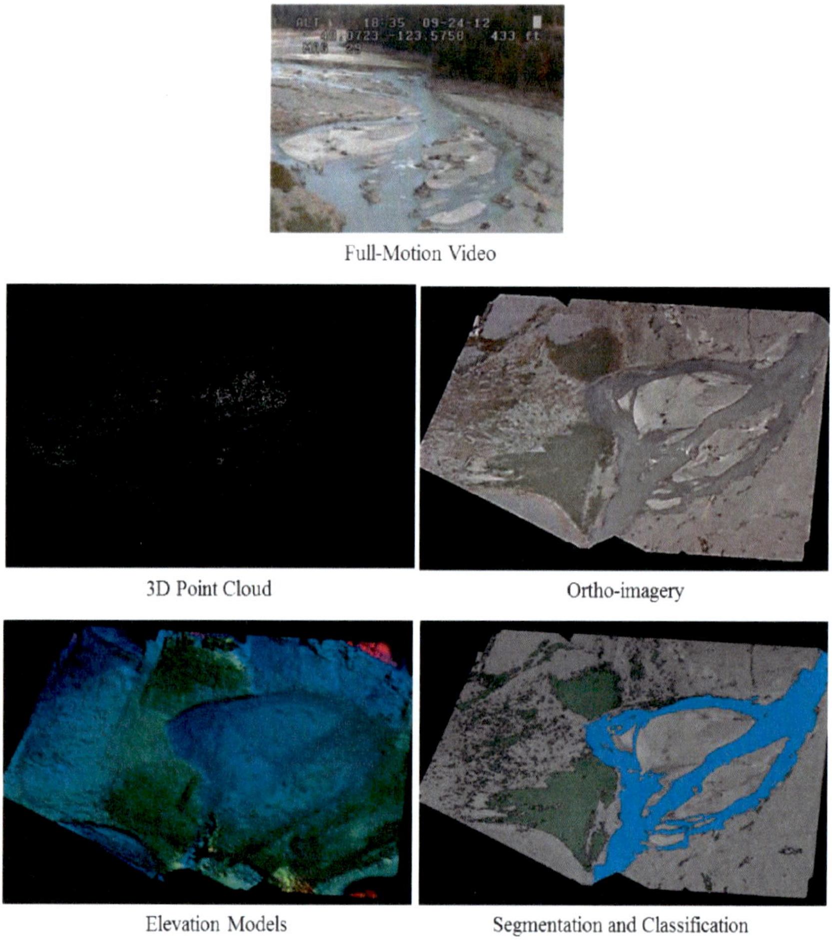


Figure 5. Examples of geospatial products that can be derived from images taken from a small unmanned aircraft.

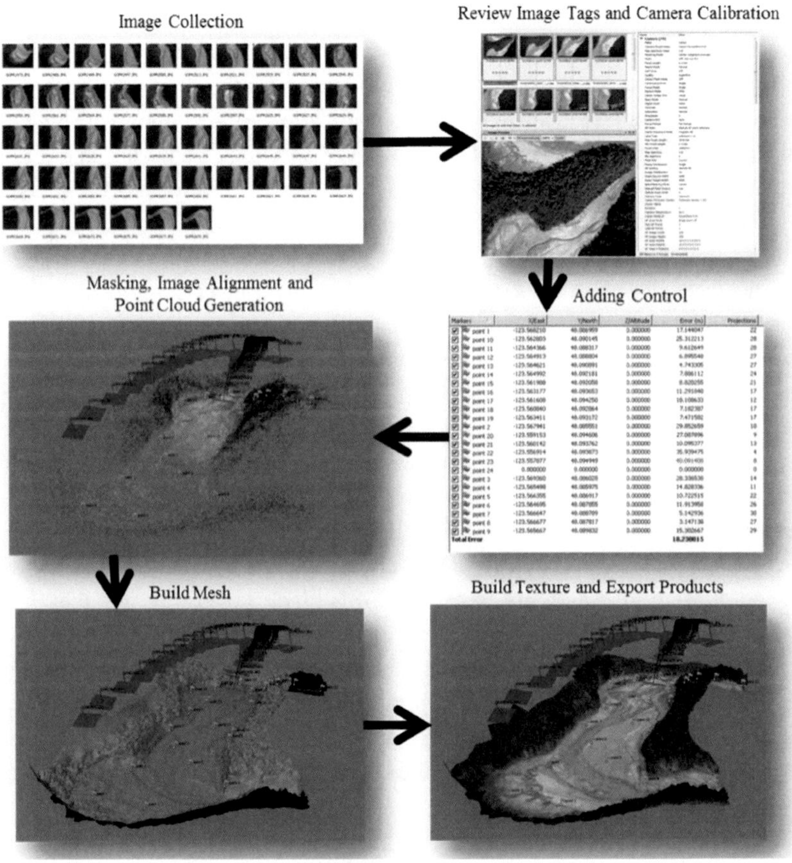


Figure 6. Diagram of the process for generating point cloud, digital elevation model, and orthophotographs from imagery taken during a small unmanned aircraft mission.

Today, high-resolution digital cameras and GPS can be coupled with cost-effective commercial off-the-shelf 3-dimensional measurement and modeling software that no longer requires knowing the distortions of each camera lens system. Such software can be used to apply a multiview solution in order to empirically model the lens distortion for any radial ground lens (effectively generating a high-quality in-place camera calibration). This advancement also means that digital single-lens reflex and point-and-shoot cameras, as well as any other camera or sensor that uses a radial ground lens, can now be deployed to acquire data.

Once the data have been preprocessed they are ready for analysis or postprocessing. The Department of the Interior has extensive experience in image analysis of airborne (optical) and satellite (optical and radar) data. This experience applies directly to the analysis of UAS data. In image analysis, preprocessed data are transformed into information, as represented by classifications, object detection and identification, and mapping. The DOI makes considerable use of ERDAS software for image analysis and ENVI software for hyperspectral analysis. PCI software, as is true also for ERDAS and ENVI, is a multisensor image-analysis package best known for its orthorectification process.

On the basis of small UAS missions that have been completed to date, most, if not all, processing of mission still imagery is done in Photoscan Professional Edition software (developed by Agisoft). Photoscan uses all available computer resources to complete advanced image calculations and render geometry, so (based on industry research and specifications proposed by one of the Photoscan developer) a custom workstation was built to use this software. A key component in this workstation was a graphics card that had high graphics processing unit (GPU) clock speeds, which are typically found on high-end desktop computers. The graphics card found on more typical desktop configurations do not have high GPU clock speeds, even if they have fast central processing units and large quantities of random-access memory. As a result, this custom-built desktop contains a graphics card with high GPU clock speeds, the latest multicore central processing units, and more than 64 gigabytes of unbuffered random-access memory (which is much less expensive than more commonly used buffered memory).

## Summary of Operational Benefits

The tangible benefits of the Department of the Interior unmanned aircraft systems operational test and evaluation program can be summed up in three words: science, safety, and savings.

Examples of these tangible benefits are as follows:

1) Small unmanned aircraft systems (sUAS) are far less disruptive to sensitive animal species than manned aircraft. They carry high-technology sensors and can transmit real-time data that can be recorded for future analysis. Many offer greater airborne persistence and launch responsiveness than manned aircraft. These characteristics enable sUAS to gather repeatable, scientifically valid observations. Better science leads to better policy decisions.

2) DOI missions often expose personnel to real safety hazards. From 1937 to 2000, 66 percent of all fatalities of DOI field biologists were related to aviation. Ground personnel in certain missions can also be replaced with sUAS, reducing the risk to its personnel and vehicles. The DOI is evaluating the role of sUAS in improving safety.
3) The cost to operate sUAS that are currently employed by the DOI is less than 10 percent of the cost to operate manned aircraft. The U.S. Fish and Wildlife Service and USGS costs for the sandhill crane sUAS mission were $2,645. Estimates to fly a similarly equipped manned aircraft ranged up to $35,000. The $15 million worth of sUAS assets currently operated by the DOI were excessed by the Department of Defense (DOD) and obtained at zero cost. Had they not been loaned to the DOI, they would have been destroyed by the DOD as excess property, depriving the American taxpayer of the full value of these assets. This "try before you buy" strategy mitigates the financial and operational risk the DOI would incur if these assets were to be acquired outright (and without equivalent support) rather than loaned with highly experienced intergovernmental operational employment and technical support. The DOI sUAS program strategy for sUAS technology (leveraging excess DOD sUAS) has resulted in substantial savings. Where sUAS can adequately replace manned aircraft, the savings to the American taxpayer are significant.

## Missions

Completed proof-of-concept missions have been key determinants of how unmanned aircraft systems (UAS) can cost effectively fill some of the DOI remote sensing data requirements (fig. 7). They have also provided the opportunity to develop and test required operational procedures and processes. This section provides more specific information and lessons learned from several missions that have completed by the USGS National Unmanned Aircraft Systems Project Office. More information about either of these sample missions, or any of the additional missions that have already been completed, is available on the Missions— page of the USGS UAS website (http://uas.usgs.gov/completedMissions.shtml).

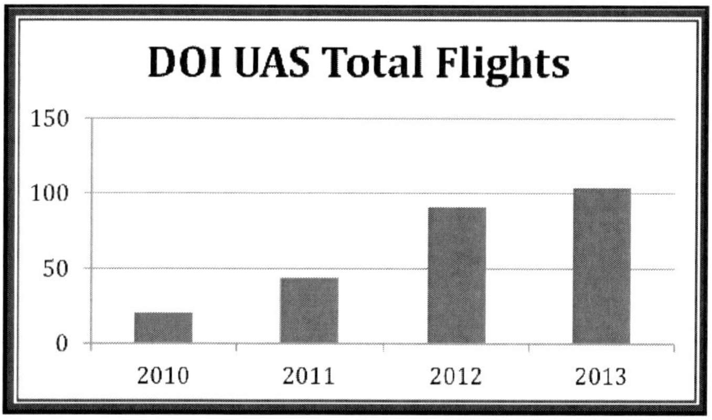


Figure 7. Chart of the total flights by Department of the Interior small unmanned aircraft in proof-of-concept missions.

Satellites, manned aircraft, or on-the-ground observations have traditionally collected the data needed for the activities detailed below, sometimes at a significant risk or cost. Even so, introducing a new technology can be challenging, because it must not only be more cost effective than traditional methods, it must also meet data specifications and operate with minimal disruption to the area being observed. Investigations performed during proof-ofconcept missions have addressed these questions and concerns and have demonstrated that UAS capabilities can cost effectively meet some remote sensing requirements and provide safe support for wildlife management, public safety, inspections, and environmental research.

## Wildlife Management

Several Department of the Interior bureaus, including the Bureau of Land Management and U.S. Fish and Wildlife Service, are mandated to manage wildlife. They do so in part by surveying threatened and endangered species, monitoring vegetation and habitat, and obtaining accurate population counts of various species.

## Sandhill Crane

The first operational small unmanned aircraft systems (sUAS) mission by the USGS National Unmanned Aircraft Systems Project Office (NUPO) took place during March 2011 at the Monte Vista National Wildlife Refuge (NWR) near Monte Vista, Colorado. In cooperation with the U.S. Fish and Wildlife Service (USFWS), the feasibility of using the sensors on a Raven sUAS to survey sandhill cranes was studied (fig. 8). The USFWS, mandated by migratory bird rules and regulation, must periodically survey the sandhill crane population. Sandhill cranes are migratory birds that travel from Texas to Idaho (or as far as Siberia) annually, and the Monte Vista NWR serves as a major stopover point during migration—and a perfect location for an accurate population count. These surveys traditionally used either fixed-wing aircraft, which can place both birds and staff at risk of mid-air collisions, or time-consuming ground-count methods, where biologists visually enumerate birds in sectional surveys. Although both of these methods produce the required information, they also introduce safety and cost concerns, which made this project ideal for a Raven proof-of-concept mission.

Figure 8. Photograph of sandhill cranes at a roosting site in Monte Vista National Wildlife Refuge, Colorado.

For this first Raven mission, the USGS NUPO established operational procedures for obtaining required approvals, such as the Certificate of Authorization (COA) issued by the Federal Aviation Administration (FAA) and for a designated radio spectrum as described earlier in this report. Work on the COA started on July 9, 2010, and approval was received 6 months later. The approved COA required that all Raven flights remain below 400 feet above ground level (AGL), remain within visual line of sight of authorized trained observers, and be aloft only between morning civil twilight and evening civil twilight. The main objective was to determine whether the Raven's sensors could identify the cranes' heat signature in enough detail to allow biologists to obtain accurate counts. Equally important was determining how the cranes would react to the Raven, especially because of concerns that the cranes would see the Raven as an eagle, a major predator, and flush or fly away when it approached. To minimize potential disruption to the cranes, the original plan was to fly the Raven at high altitudes or at night when the cranes were at roost. However, these plans were revised when the approved COA specified altitude limitations and day flights only. Because the vast majority of cranes depart the roost at or shortly after sunrise, Raven flights were planned for the approximate 30-minute interval between onset of civil twilight and sunrise.

The first Raven flight flew about 800 meters to the roosting site, where two certified observers were stationed. During this flight the cranes did not visibly react, putting concerns to rest. The pilot operator reported that the cranes were clearly visible on the Raven's front-looking infrared camera. Several additional flight passes were made at altitudes ranging from 75 feet to 300 feet AGL before the Raven was returned to the launch site and landed. After the results of these completed flights were evaluated, flight plans for the next day were set to use a modified down-looking thermal camera and to fly an area 600  400 meters by using parallel line transects at 200 feet AGL to cover the complete roost site. On the basis of these decisions, flight preparation used the FalconView mission planning software, available on the Raven, to set waypoints that created an east-west transect grid over the entire roosting area, with overlapping flight paths, at an altitude of 200 feet AGL (fig. 9).

The next day the Raven, with its nadir infrared sensor, was flown in its autonomous flight mode; it successfully completed all transect runs within 13 minutes. The waypoints were then adjusted and the Raven flew seven more runs at higher altitudes, 300 feet and 400 feet AGL. Each of these flights lasted

24 minutes and made a total of 13 passes over the roost. This mission resulted in a total combined flight time of 4 hours, all without disturbing the cranes.

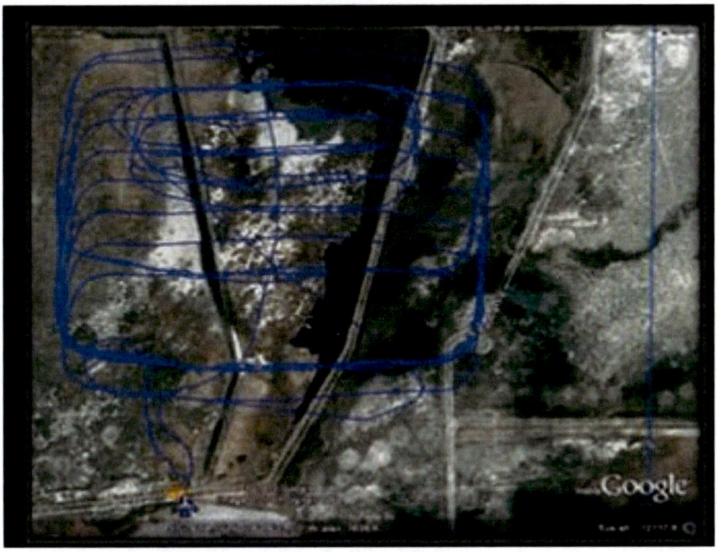


Figure 9. Image of a preprogrammed Raven flight path over the Monte Vista National Wildlife Refuge, Colorado. Blue lines represent the programmed flight lines for the aerial survey. Courtesy of U.S. Geological Survey, U.S. Fish and Wildlife Service, and Google.

These flights demonstrated that the Raven sUAS could be flown without disruption to the cranes and that the sensor payloads could identify their heat signature. One question remained: whether these data would provide accurate population counts. The counts were first derived from imagery captured by the Raven sensors. Mosaicking software processed the video into a single image of the complete roosting area (fig. 10). Next, the easily identified heat signatures were counted and a Raven-imagery population count was derived. This population count was then compared to a count already established by the USFWS during an earlier ground count that located the cranes within their roosting area and obtained an "observed"' population count. The ground-based and UAS data derived population counts were 4.6 percent different, which was determined to be acceptable.

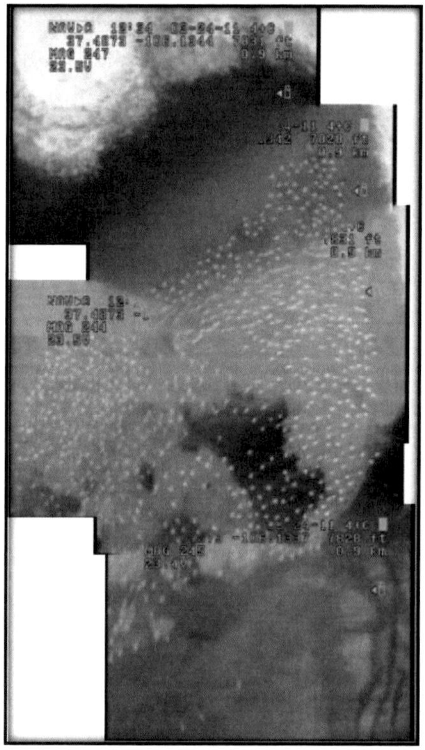

Figure 10. Image mosaic of a sandhill crane roosting area at Monte Vista National Wildlife Refuge, Colorado. Mosaic created from still-frame images captured by an infrared video camera mounted on a Raven. White dots are sandhill crane heat signatures.

These results established that sUAS can make safe, cost-effective, and accurate population counts of sandhill cranes. Dave Sharp, a retired USFWS biologist who has worked with sandhill cranes for more than 25 years, said "I was skeptical about this process at first, but it has exceeded my wildest expectations." When asked if Raven video could be used to validate their counts, another biologist said that it was the other way around—his count will be used to validate the Raven's: "The Raven found birds that we did not know were there." The Refuge Manager, Floyd Truetken, said "I would like Raven and the team to come back to my refuge next year and they can come and train any time. I would definitely recommend them to other refuge managers and biologists." The scientists agreed that this technology will revolutionize the way wildlife counts and tracking are conducted in the future.

After successful completion of this first mission, the FAA approved USGS NUPO for night flights during later Raven missions at this NWR. These additional missions, including night flights, took place in November 2011, March 2012, and March 2013. They further established that Raven imagery provides acceptable population counts without disrupting the cranes, especially when flown at night when cranes are at rest in roosting areas.

This first sUAS mission in March 2011 cost just $2,645, compared to similar fixed-wing manned aircraft surveys that cost from $4,310 up to $35,000 if contracted privately. These results helped to establish that using sUAS to perform population counts for sandhill cranes produces accurate results and is both safe and cost effective.

### *Pygmy Rabbit Habitat*

In June and July 2011, the USGS National Unmanned Aircraft Systems Project Office partnered with Boise State University and the University of Idaho to determine if small unmanned aircraft systems (sUAS) could help gather data on the landscape habitat of pygmy rabbits (fig. 11). Pygmy rabbits dig extensive burrow systems in areas with large sagebrush, particularly mature stands, and loss of habitat is a direct threat to the species. These rabbits are an important food source for many other animals in the area; birds and other mammals are primary predators. However, the rabbit's habitat has become increasingly fragmented by development, agriculture, fire, and rangeland projects such as those that replace big sagebrush with bunchgrasses.

Figure 11. Photograph of a pygmy rabbit held by a Boise State University student. Courtesy of Boise State University.

The goal of this mission was to determine if Raven sensors could remotely sense the quality of vegetation cover (for example, concealment or transparency), diet (for example, nutrients and chemical defenses), and temperatures throughout the landscape, and to overlay these attributes with remotely sensed movement and habitat use by wildlife. These data could then be used to determine the fitness of pygmy rabbits and potentially aid agencies that must decide how to use land for habitat conservation.

Several low-altitude Raven flights over the south-central desert in Idaho captured high-resolution electro-optical data. Later, these data were incorporated into a base layer of rectified imagery that could be used to analyze pygmy rabbit habitats (figs. 12 and 13). This rectified-imagery layer was then used to develop and field test spatially explicit models (for example, mixed spatial regression models and hierarchical Bayesian models) depicting how vegetation cover, quality of diet, and temperatures influence habitat use by pygmy rabbits.

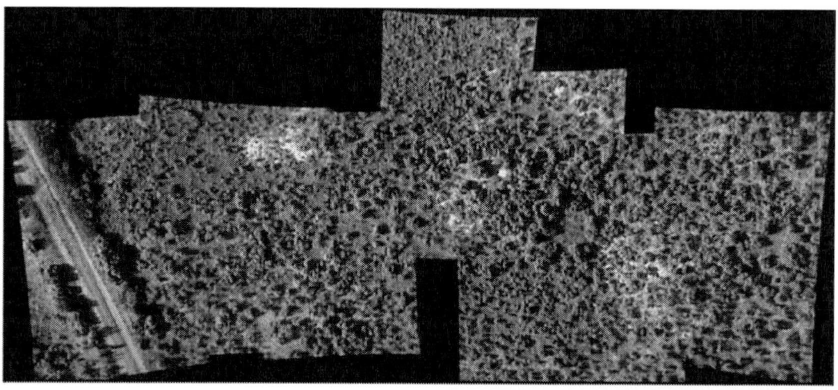

Figure 12. Image mosaic of potential pygmy rabbit habitat in Idaho. Mosaic created with still-frame images taken by an electro-optical video camera on a Raven.

### *Greater Sage-Grouse*

The Colorado Parks and Wildlife (CPW) manages and conserves the State's wildlife, including the greater sage-grouse, which is listed by the U.S. Fish and Wildlife Service as a species warranting, but precluded for, Endangered and Threatened Species status. To best manage greater sage-grouse, the location of breeding sites (leks) and accurate population counts for each lek are needed. For example, the CPW uses high male counts from all of the leks in northwestern Colorado to assess population trends, but locating new leks can involve hours of searching, mostly on foot, on large expanses of land with rolling topography that can limit vantage points. In Grand County an established lek has eluded detection, even after extensive ground searches in areas where greater sage-grouse are seen and flushed not only during the breeding season but also throughout the rest of the year.

Figure 13. Larger scale of the image mosaic shown in figure 12.

The CPW, in cooperation with the Bureau of Land Management and the USGS, initiated a proof-of-concept mission to determine if the Raven and its sensors was an option for safely and effectively detecting and counting sage-grouse on lek sites. This mission aimed to answer the following questions:

- How would the greater sage-grouse respond to Raven small unmanned aircraft systems (sUAS) flights?
- Will visible wavelength or thermal infrared sensors better detect greater sage-grouse?
- What is the optimum flight altitude for detecting these birds but not disturbing them?
- Can the sensor detect greater sage-grouse obscured by sagebrush? and
- Can the sensor distinguish greater sage-grouse males from females?

In addition, it was hoped that use of the Raven would allow CPW to search large expanses of open sagebrush habitats for breeding activity and identify new leks.

This mission was successfully flown in April 2013 near four leks in Middle Park, Colorado. A thermal infrared video camera acquired data during the early twilight hours and a GoPro Hero2 camera was used during predawn to capture electro-optical (EO) images. During these flights the sage-grouse, similar to the sandhill cranes, were not disturbed by the Raven and, in fact, remained on the leks and continued breeding activities. Both the thermal infrared and EO cameras were able to detect the birds from altitudes of 100–200 feet above ground level (figs. 14 and 15). The thermal camera detected the birds only as blurred objects, making distinguishing male from female impossible, and even the higher resolution EO camera had a hard time distinguishing the smaller female sage-grouse or detecting birds hiding under sagebrush. However, the data generated accurate population counts either manually through visual interpretation, or partly automated with software designed for feature recognition. It also provided the beginning of a historical archive that could be used to build sequential time-series records for future research.

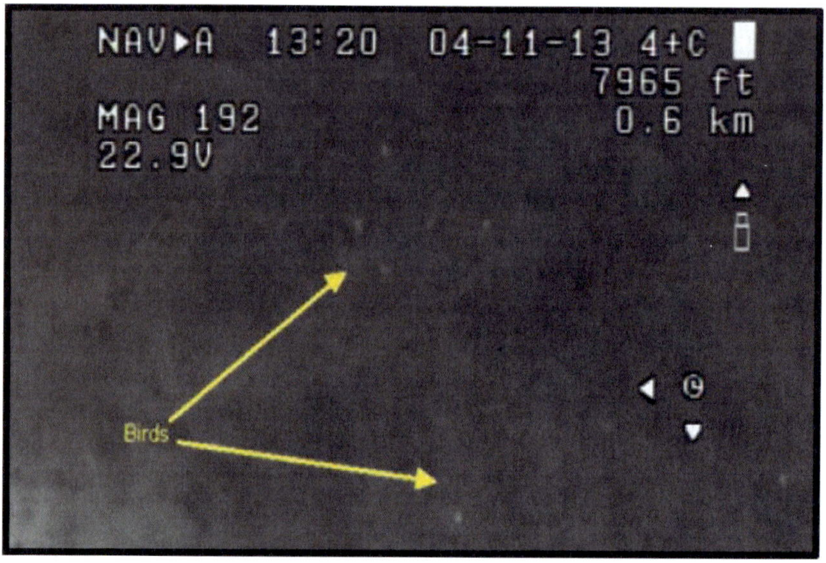


Figure 14. Thermal image in Middle Park, Colorado. Image taken at 200 feet above ground level by a video camera on a Raven. Yellow arrows point to thermal heat signatures of greater sage-grouse.

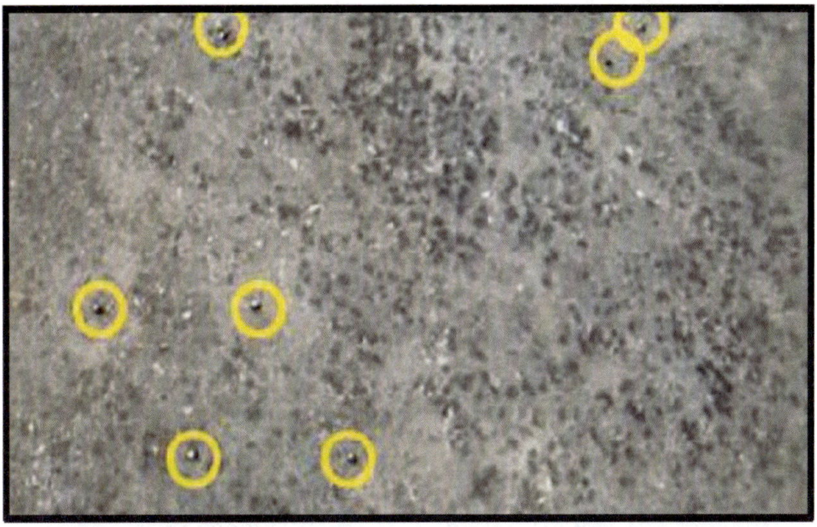

Figure 15. Color electro-optical still-frame image in Middle Park, Colorado. Image taken from 160 feet above ground level by a video camera on a Raven. Yellow circles indicate greater sage-grouse.

Although the current set of Raven sensors was unable to identify birds under sagebrush or distinguish male from female, it did produce images useful for population counts. Overall, this mission proved that sUAS provided a low-cost, high-resolution, and mobile platform for determining sage-grouse population counts. Better thermal and EO cameras, which are currently being developed, would greatly aide in meeting mission objectives.

***Waterfowl Surveys***

The Ruby Lake National Wildlife Refuge, which is managed by the U.S. Fish and Wildlife Service (USFWS), is located in the Great Basin region in Nevada, a high desert at an elevation of nearly 6,000 feet. Its marshes, meadows, grasslands, and shrub-steppe upland provide habitats for thousands of nesting and migrating waterfowl. This unique wetland in the high desert is visited by more than 220 species of birds and migratory waterfowl.

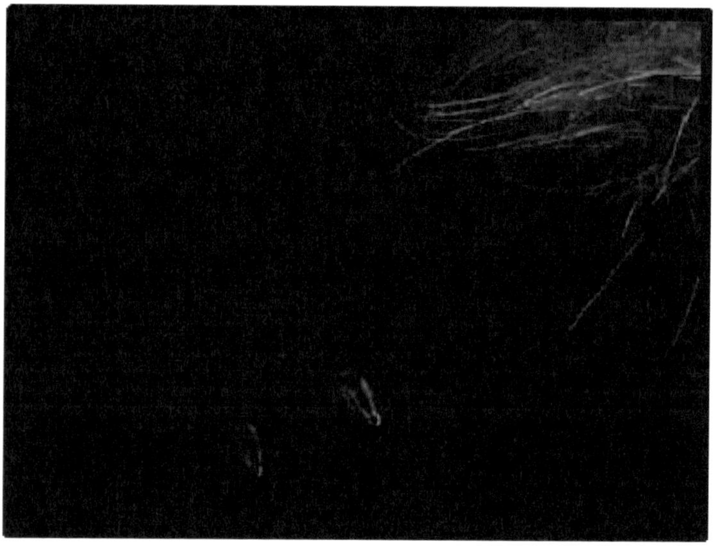

Figure 16. Image of mallard ducks in Ruby Lake National Wildlife Refuge, Nevada. Image taken by a Canon SX230 camera on a 1-Hawk.

In November 2013 the National Unmanned Aircraft Systems Project Office, in cooperation with the USFWS, completed a mission to determine if small unmanned aircraft systems (sUAS) sensors could be used to count general waterfowl populations and whether they could distinguish individual species (fig. 16). Another mission objective was to determine whether this data could also be used to map nesting areas and identify vegetative health; both types of information vital in supporting optimal refuge management.

During this mission, data were collected at several flight heights and by using several different sensors, such as HD GoPro Hero3 video, Canon SX230 natural color imagery at various zoom levels, and Canon SX260 color infrared (fig. 17).

Data analysis after the mission determined that high-resolution data captured from about 100 feet above ground level by using zoomed camera settings show the best potential for counting waterfowl and identifying individual species.

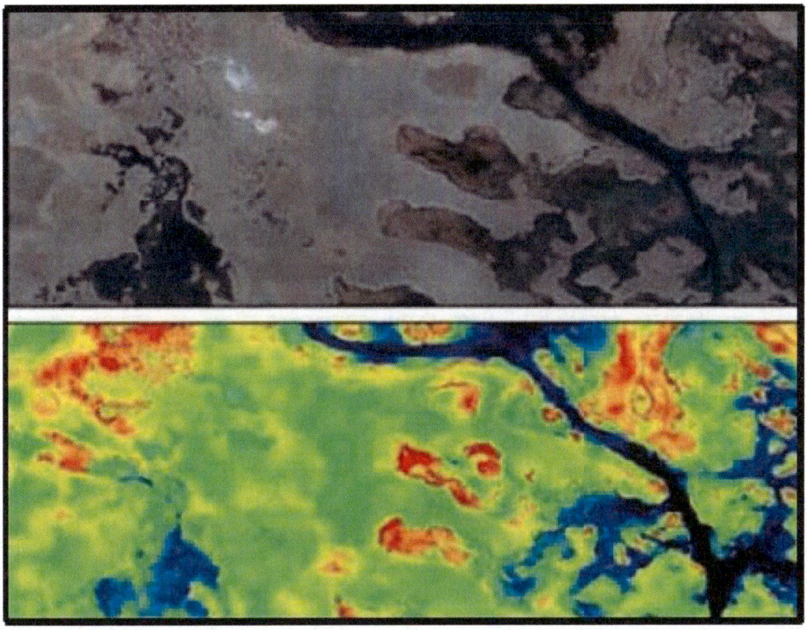

Figure 17. Color infrared image mosaic of the Ruby Lake National Wildlife Refuge, Nevada (top). Mosaic derived from images taken by a Canon SX260 on a 1-Hawk. Normalized-difference vegetation index derived from the color infrared image mosaic (bottom).

## Public Safety

Several bureaus in the Department of the Interior ensure public safety by providing emergency response during any natural hazard event on the public lands they manage. These events, such as volcanic eruptions or wildland fires, require rapid response and timely access to current data. The ability to use unmanned aircraft for proof-of-concept public safety or rapid-response missions have been hampered by the time-consuming preflight approval requirements. Changes in these requirements, such as those in the memorandum of agreement between the Department of the Interior and the Federal Aviation Administration, should soon make these missions more feasible. See the Rapid-Response Missions section, found later in this document, for additional information.

***Prescribed Fire***

In February 2011, a Raven mission was flown during a prescribed burn at Eglin Air Force Base, Florida by the USGS National Unmanned Aircraft Systems Program Office in cooperation with the U.S. Forest Service (fig. 18). This burn was part of the Prescribed Fire Combustion and Atmospheric Dynamics Research Experiment coordinated by the Air Force and several fire researchers from throughout the nation. It provided the first opportunity to operate Raven small unmanned aircraft systems (sUAS) in the United States. As a plus, because the Raven was flown within military air space, no additional approvals were required.

Figure 18. Unmanned aircraft systems base station at a prescribed burn at Eglin Air Force Base, Florida. Smoke in the background was produced by the fire.

This mission tested the operation of sUAS technology and its various sensor payloads, particularly streaming live video to a command center to support fire management and operations. During the Raven's flight, one of three different sUAS platforms that were flown, both the thermal infrared and electro-optical natural color sensors were used, and the associated live video was stored for later analysis (fig. 19).

The Raven successfully demonstrated its live-streaming capability and its image data-collection ability, which allows storage of the live imagery for use later in research or as a historical record. During the Raven flight, technical problems interfered with some communications between the vehicles and command center equipment. However, these problems allowed the crew to demonstrate how to handle loss-of-link problems and complete a mission.

Figure 19. Thermal infrared image of a prescribed burn at Eglin Air Force Base, Florida. Image taken from a Raven flying about 500 feet above ground level. White region is hot.

## Inspections

The Department of the Interior must ensure that public assets such as forests, parks, and active mine sites are maintained and operated safely for employees and visitors, and with minimal degradation to associated ecosystems. This requires bureaus to perform periodic inspections of vast areas, including all of the national parks in the continental United States, Alaska, and Hawaii.

### *Contour Surface Mine*

Mine site inspections are necessary to monitor a range of topics: water quality, hazardous conditions, terrain, wildlife habitats, land use after mining, and safety of cultural features. For instance, each coal mine is required to install drainage-control structures internally and around the perimeter to control runoff from the disturbed area and to prevent flooding. Verifying this requirement requires vast amounts of time for State and Federal inspectors, especially in states such as West Virginia, which contains more than 1,800 coal mines. West Virginia also has many underground mine fires about which very little is known, sometimes not even the location and extent of the fire. But because locating and inspecting each of these underground fires would require traversing rough terrain hazardous to both the inspectors and their vehicles, these fires generally go undetected.

During the interval 2011–2013, three small unmanned aircraft (sUAS) missions were flown at the Powellton surface mine, operated by the National Unmanned Aircraft Systems Project Office (NUPO), in coordination with the Office of Surface Mining Reclamation and Enforcement. The Powellton mine is a contour surface mine that has several miles of installed drainage and sediment ditches and two known underground mine fires, which made it ideal for testing the utility of sUAS sensors for supporting inspections of mine drainage and underground fires (figs. 20 and 21). The first mission, in November 2011, used Raven sensors to inspect the area, but the next two missions (in 2012 and 2013) used a T-Hawk. The T-Hawks ability to hover and stare, and the ease with which cameras could be mounted for onboard data capture, quickly made it the platform of choice. Additionally, as compared with the Raven, the T-Hawk was more maneuverable in tighter areas with steep terrain.

Figure 20. Image of sediment drainage ditches at the Edwight surface mine in West Virginia. Image taken by a GoPro Hero camera on a T-Hawk.

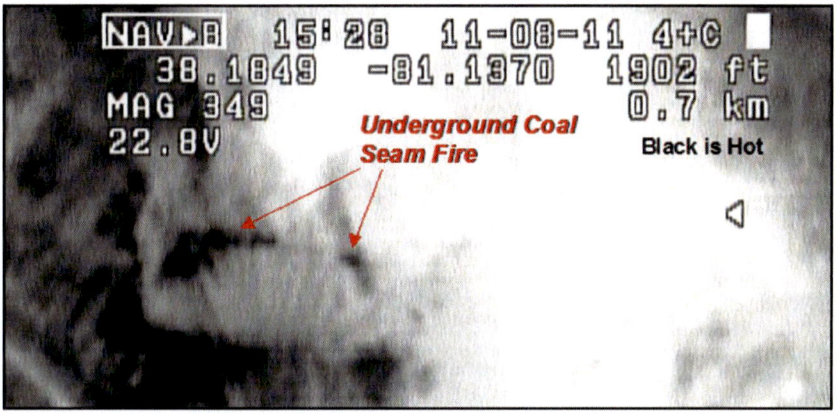


Figure 21. Thermal infrared image of underground coal-seam fires in West Virginia. Image taken by an infrared video camera on a Raven. Black region is hot.

Figure 22. Oblique-angle image of sediment dams at the Republic surface mine in West Virginia. Image taken by a GoPro Hero camera on a 1-Hawk.

Data acquired during these missions include video files from external camera sensors mounted onboard the platform and the full-motion video used to pilot the platform. High-resolution still-frame image data were also captured by the external camera sensors, such as the Canon and the GoPro, at both oblique and vertical angles. Later analysis determined that the imagery taken at oblique angles was as valuable as vertical imagery for performing visual inspections to insure compliance (fig. 22). After the missions photogrammetric software was used to generate 3-dimensional models of the mining areas in order to make perimeter and volume measurements.

These successful missions at West Virginia coal mines proved that sUAS technology was a uniquely efficient new tool to aid inspectors in monitoring the vast areas of surface mines they must oversee annually. The imagery also effectively captured coal-seam fire locations and extent. In addition, valuable secondary products, such as 3-dimensional modeling, reclamation ground measurements, time-series investigations, color infrared images for seep detection, and georectified base images, were derived from the low-altitude imagery (fig. 23).

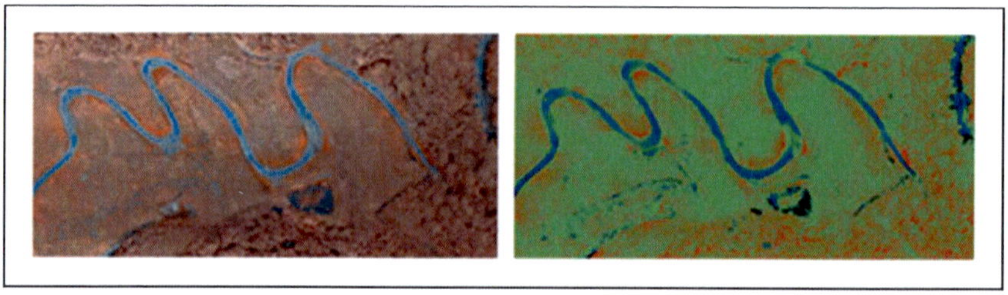

Figure 23. Color infrared images of a mining impoundment in West Virginia. Images taken by a Canon SX260 camera on a 1-Hawk. Image on the left is color-infrared and image on the right is NDVI.

### *Resource and Vegetative Species*

Haleakala National Park, on Maui in the Hawaiian Islands, consists of 30,183 acres of public land in remote and rough terrain that ranges from sea level to 10,023 feet (fig. 24). The Haleakala Resource Management Division preserves and restores resources in this national park through activities such as surveying invasive plants, monitoring wildlife, and maintaining a 50- mile-long fence that protects the park from feral pigs and goats. Work crews can spend hours in rough terrain simply trying to find the location of invasive plants, such as *Hedychium gardnerianum* (ginger lily) and *Miconia calvescens* (bush currant), and inspecting the barrier fence for holes, damage, and other needed maintenance.

Figure 24. Photograph of a Raven being launched from a mission site in Haleakala National Park, Hawaii.

To help improve staff safety, the USGS National Unmanned Aircraft Systems Project Office, in coordination with the National Park Service, flew a proof-of-concept mission in May 2012 for mapping invasive plants and inspecting fence lines in inaccessible areas. This mission was the first to directly benefit from the Raven's real-time full-motion video, which displayed the fence line while the Raven was in flight allowing coordinates to be noted for areas requiring repair or further inspection. The Raven was also small and slow enough to pose little harm to native birds, and it emitted much less noise than a typical manned aircraft, thus limiting disruptions to wildlife.

This mission demonstrated the feasibility of using small unmanned aircraft for safe, real-time inspection of items such as fence lines in rugged terrain. Moreover, its base imagery data could be used for identifying invasive plants and monitoring vegetation health. This mission also showcased the Raven's ability to operate in harsh conditions at high elevations and in very strong winds.

### *Abandoned Mine Lands*

Abandoned mines can pose serious health and safety hazards, such as landslides, erosion, surface instability, and burning coal refuse, which is why the government is required to inspect abandoned mine sites and identify any needed remediation. State and Federal agencies spend large amounts of time and money driving to remote sites through rough terrain in order to inspect those sites. Therefore, in September 2012 a proof-of-concept mission was launched to determine if Raven sensors could be used to identify features of an abandoned mine, such as portals or other dangerous openings. Using mine-site data acquired by a Raven would markedly decrease the time and dollars expended on these inspections and would also provide data to help inspectors determine the best course for any needed remediation.

This mission took place at the Coal Basin mining operation, Pitkin County, Colorado (fig. 25); it was operated by the National Unmanned Aircraft Systems Project Office, in cooperation with the Office of Surface Mining and Reclamation. Mine facilities were located in a topographic basin of 236 acres on the divide that separates the North Fork of the Gunnison River watershed from the Crystal River watershed. Drainage from the east side of Huntsman Ridge flows through several tributaries into Coal Creek. All mine portals and surface facilities in the basin are located in Pitkin County, primarily within the White River National Forest. The underground mine areas extended to the west under Huntsman Ridge into Gunnison County.

Figure 25. Image of a drainage pond at the Coal Basin mine in Colorado. Image taken by a GoPro Hero camera on a Raven.

Mining of Coal Basin deposits began in 1895 and continued until 1908; much later, in 1956, Mid-Continent Resources began producing coking coal there. The Coal Basin mine consists of five adjacent underground mines, a rock tunnel entry, a preparation plant, two coal waste piles, one development waste pile, an extensive road system, and numerous ancillary facilities. Five drift mines, driven from the outcrop down-dip through the western flank of Coal Basin and under Huntsman Ridge, were all interconnected. Coal, men, and equipment were brought to the surface through one bore of the rock tunnel. Additionally, the Coal Basin area contains a precipitous change in elevation: the lowest point was the coal preparation plant at 8,000 feet, whereas the highest point is along Huntsman Ridge at 11,852 feet. As a result, temperature, precipitation, and wind conditions are quite variable throughout the disturbed area.

During this mission a GoPro Hero 2 camera mounted on the Raven was first used— making this the first time that a commercially available camera was used by the USGS National Unmanned Aircraft Systems Project Office. Later, these data were used to provide overlapping stereo images as input for 3-dimensional photogrammetric modeling software, to generate realistic 3-dimensional models (fig. 26).

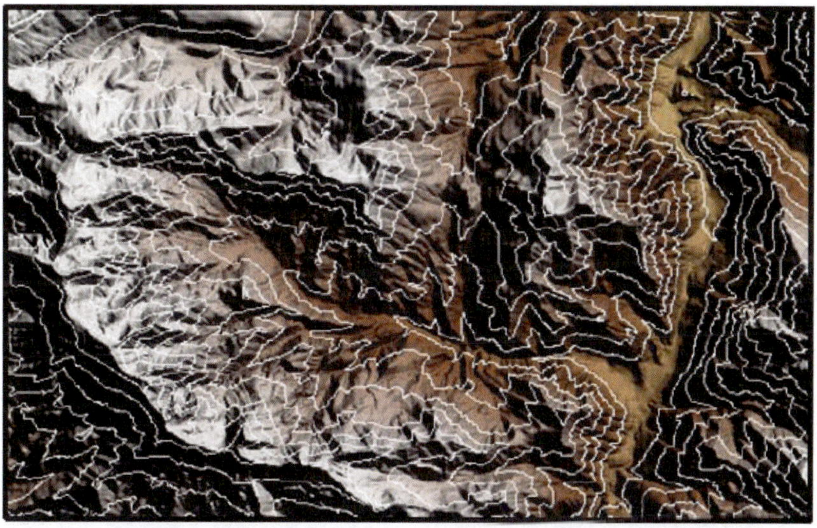

Figure 26. Picture of an elevation model with 500-foot contours in Pitkin County Colorado. Model derived from an image taken by a GoPro Hero on a Raven.

Overall, this mission demonstrated the value of mounting commercially available sensors on small unmanned aircraft platforms to assist in the difficult job of monitoring and inspecting the vast abandoned mine land areas. Also showcased during this mission was the Raven's ability to fly at elevations above 10,000 feet, which supports monitoring the many high-elevation Department of the Interior lands in the western United States.

### *Abandoned Solid Waste*

In March 2013, the USGS National Unmanned Aircraft Systems Project Office, in cooperation with the National Park Service Mojave National Preserve, flew proof-of-concept small unmanned aircraft systems (sUAS) missions in an isolated part of southeastern San Bernardino County (fig. 27). The primary objective was to determine if this new technology could provide cost-effective, high-resolution aerial imagery of the preserve. This imagery in turn would provide a historical record of abandoned solid waste, aid the difficult task of identifying abandoned materials, determine historical significance, and support cleanup. Freely available satellite imagery does not meet the current resolution requirements, and unmaintained two-track roads hamper ground efforts to access sites.

Figure 27. Image of abandoned solid waste in Mojave National Preserve, California. Image taken by a GoPro Hero camera on a Raven.

The passage of the California Desert Protection Act (CDPA) of 1994 (Public Law 103-433) created the Mojave National Preserve and placed it under the management jurisdiction of the National Park Service. Until then, the land had been used for Native American life, mining, grazing and, during the 1920s and 1930s, homesteading. Most homesteads were developed in Lanfair Valley, a fairly flat, broadly oval expanse of about 150 square miles (120,000 acres) of Joshua tree– creosote shrub habitat. Homesteading converted much of the publicly owned valley to private ownership, but drought and hardship during the 1930s triggered out-migration of homesteaders, whose land parcels were subsequently handed down through families and divided or sold. After this time the land, if used at all, was generally used for recreational purposes and not always well maintained. By 1994 when the CDPA was passed, roughly half of Lanfair Valley was a patchwork of private and public land. Most parcels were 40 acres or smaller, and many contained accumulated abandoned solid materials, some with historic value. The newly established CDPA dictated that private land within the preserve should be acquired, if willing sellers exist and acquisition funds are donated from private sources.

Today, a primary land-management need on the 1.6 million acres in the preserve is to locate abandoned nonhistoric material, much of which is in Lanfair Valley, evaluate its significance, and determine if cleanup is required. Historic materials are generally left at the site, providing discovery and recreational opportunities for the 20 million people who live within a 4-hour drive of the preserve.

Low-altitude imagery was acquired over four areas of approximately 1 square mile by using a GoPro Hero 2 camera mounted on a Raven and a Canon SX260 camera mounted on a T-Hawk (fig. 28). Data viewed during the flights identified abandoned material in some locations— primarily in ditches or low valleys—that would be very difficult to see from higher altitudes.

This mission successfully demonstrated that sUAS can make good low-altitude reconnaissance and monitoring tools, because image data could provide, either in real time or during postprocessing, not only identification of abandoned materials but also exact coordinates. This coordinate information allows the inspectors to easily navigate directly to the required inspection sites.

An unexpected benefit was the discovery that this high-resolution imagery could be used to distinguish various vegetation types, such as Joshua trees (fig. 29). Image data with this degree of resolution can be used with automated extraction techniques to create valuable vegetation inventories, establishing another potential use for the data.

Figure 28. Photograph of a 1-Hawk launch in the Mojave National Preserve, California.

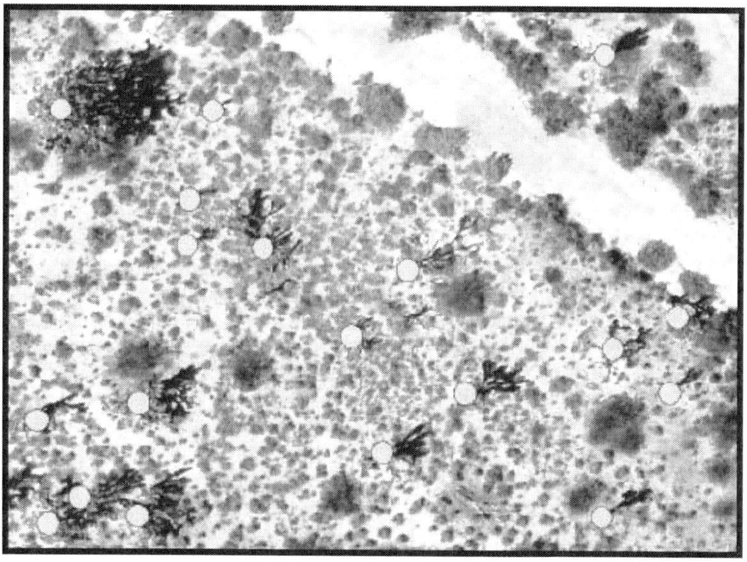

Figure 29. Image of Joshua trees (dots) in the Mojave National Preserve, California. Image taken by a Canon SX260 camera on a 1-Hawk.

## Research

Department of the Interior bureaus support scientific investigations that analyze various types of Earth surface processes and the effect they can have on society, resources, and economic development. This focus includes water studies, because society depends on fresh, reliable water supplies, as do diverse and fragile ecosystems. Information about water—use, flow, quality, and availability forecasting—are important to best meet water needs and protect the environment.

### *Groundwater Discharge*

Red Rock Lakes National Wildlife Refuge, Montana, contains one of the few marshland wilderness areas in the country, and it hosts rare trumpeter swans. Groundwater discharge to Upper and Lower Red Rock Lakes and the wetlands in-between contributes a large share of the marshland's water budget. Understanding the volume, distribution, and temporal variability of groundwater discharge is the basis for management of the natural resources of this refuge and wilderness area. However, the volume and distribution of groundwater discharge remain unknown.

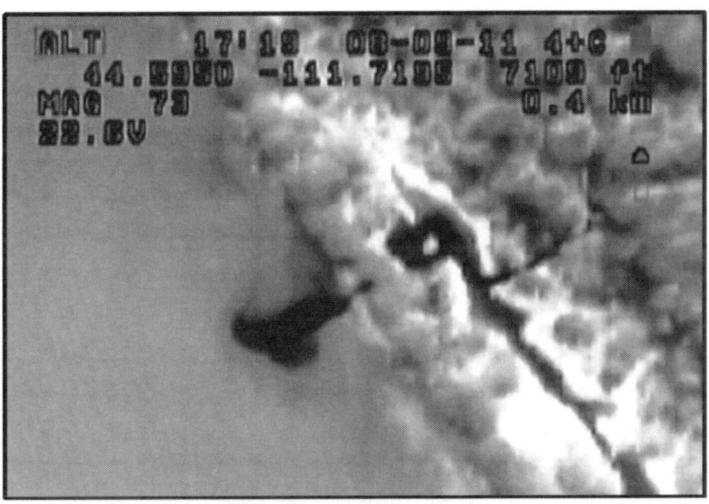


Figure 30. Thermal image showing temperature differences depicting groundwater discharge in lake water in Red Rock Lakes National Wildlife Refuge, Montana. Image taken by an infrared video camera on a Raven.

During middle to late summer, discharge water is about 15 degrees C colder than shallow lake water. In August 2011, USGS flew a Raven to determine if its thermal sensors could identify areas of groundwater discharge (fig. 30) based on this temperature difference. It was hoped that even these uncalibrated thermal sensors could identify relative temperature differences adequate for mapping the extents of the discharge areas within the lakes or near the shoreline. Areas of focused groundwater discharge could then be identified, after which measurements of rates of discharge will allow quantification of this focused groundwater discharge.

The Raven's thermal data distinguished several springs in the near-shore margins of Upper Red Rock Lake. Some of these springs discharged on land adjacent to the lake and the spring water then flowed overland into the lake, while others discharged in shallow water within several hundred meters of the shoreline. Some springs were readily visible as distinct holes in the lakebed, whereas others were harder to identify.

Relative temperature differences, particularly in water bodies, turned out to be fairly easy for the thermal infrared sensors on the Raven to detect. Thus, this technology is useful for mapping groundwater discharge point and distribution. Moreover, time-series studies could be used to display the perennial or intermittent nature of the flow.

### *Shoreline Erosion*

The Lower Brule Sioux Tribe Reservation, home to approximately 600 Native Americans, is located in central South Dakota on the western side of the Missouri River. Shorelines of the Missouri River and two lakes behind dams, Lake Sharpe and Lake Francis Case, form the northern and eastern boundary of the reservation. These shores are prone to rapid erosion, as much as 8 feet per year. This erosion threatens the many ancestral sites located along the Missouri River, as well as agriculture and recreational activities that contribute a large part of the reservation's economy. To better evaluate the effects of this erosion, the tribe's Environmental Protection Office in South Dakota established a 2-year study of a site that contains a 7-mile stretch of Missouri River shoreline and the intake location of the reservation's rural water plant. Direct access to the shoreline is hampered by lack of roads, crumbling riverbanks, and shallow water unsuitable for motorboat traffic (fig. 31). In one effort to address these accessibility problems, the tribe's Environmental Protection Office asked the USGS National Unmanned Aircraft Systems Project Office to determine if Raven sensors could help monitor this shoreline.

Figure 31. Photograph of a rapidly eroding section of Missouri River shoreline in South Dakota.

During missions in August of 2011 and 2012 the Raven safely and easily accessed this remote shoreline; it was hand launched from land away from the shoreline or from a motorboat offshore. The Raven captured video imagery along the shoreline with its original full-motion video analog sensor, which was the only small unmanned aircraft sensor available at the time. This relatively low-resolution video was later used to create still-frame photographs that were georeferenced with existing base imagery. As the final step, the rectified images were mosaicked to create a seamless shoreline image (fig. 32). This process was repeated with the data acquired during flights in 2012, resulting in 2 years of orthorectified and oblique imagery over the same river bank erosion areas. Such time series can be used to detect change, evaluate rates of erosion, and aid land-management decisions.

This successful mission demonstrated the benefits of this technology for providing a safe and economical alternative method for monitoring inaccessible or dangerous shorelines.

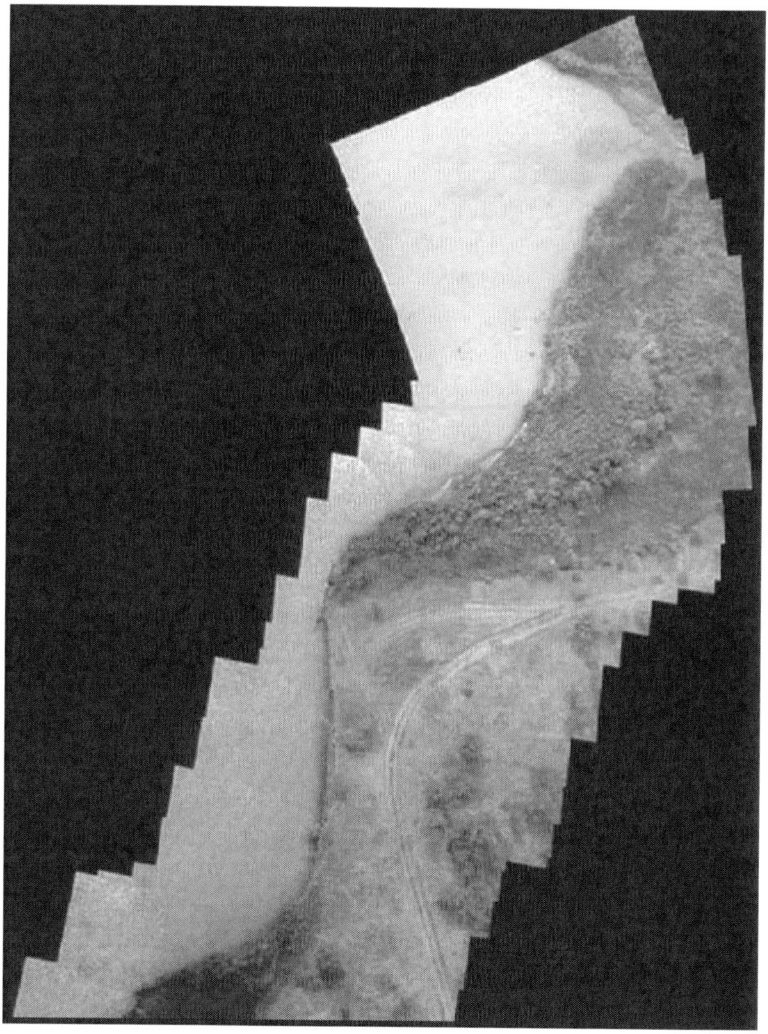

Figure 32. Image mosaic of a section of Missouri shoreline in South Dakota. Mosaic created from rectified images taken by an electro-optical video camera on a Raven.

***River Restoration***

The Elwha River Restoration Project, led by the National Park Service, was established to produce a scientifically sound technical narrative describing what happened to the fish, reservoir sediment, reservoir topography, and vegetation during and following dam removal in the State of Washington. This restoration project is unique, because it represents the largest controlled release of sediment in the history of North America, and it encompasses an entire watershed (fig. 33). The entire area contains 320 square miles, from the Olympic Mountains to the Strait of Juan de Fuca, of predominantly pristine wilderness that is also historically home to all runs of Pacific Northwest salmon. Fully understanding the physical and biological responses of this river system to dam removal will provide invaluable information needed for other dam removal river restoration projects, as well as for any dam operations undergoing Federal Energy Regulatory Commission relicensing.

Figure 33. Image of the dam removal site at Lake Mills, Washington. Image taken by a GoPro Hero camera on a Raven.

In June and September 2012, small unmanned aircraft (sUAS) missions were flown along the Elwha River to aid this restoration effort, as a collaboration of the USGS, the Bureau of Reclamation, and the National Park Service. The goal was to use repeated sUAS surveys to acquire data that, once orthorectified, would provide information about the rates and patterns of change during this large-scale river restoration. Because this dam removal is the first on such a large scale, it was important to monitor how rapidly sediment volumes eroded from the reservoir and were deposited downstream (fig. 34). The remotely sensed data collected by the sUAS can be orthorectified and used to generate elevation models to monitor changes in the reservoirs and river channel, especially where redeposited sediment may affect salmon habitat and the height of flood stage.

Figure 34. Image of the Lake Aldwell area in Washington. Image taken from 200 feet above ground level by a GoPro Hero camera on a Raven.

After they were acquired, imagery data were used in a systematic process of geometric correction and 3-dimensional reconstruction to generate a river model over the Lake Aldwell area. Data-processing steps included selecting, loading, and aligning the images; creating a point cloud with horizontal and vertical derivations; elevation modeling; draping of the images; and export of the geographic products (fig. 35). This processing produced an orthorectified imagery base of the reservoir, showing the restored river basin with the rapid change in sediment movement and infill. Combining this base imagery with other remotely sensed data types provides a multilayered historical collection of data over the area that can be used to evaluate sediment movements. Documenting this information will also benefit future efforts, because what is learned from the Elwha River changes can be used as a guide during other dam removal projects in the Pacific Northwest and nationwide.

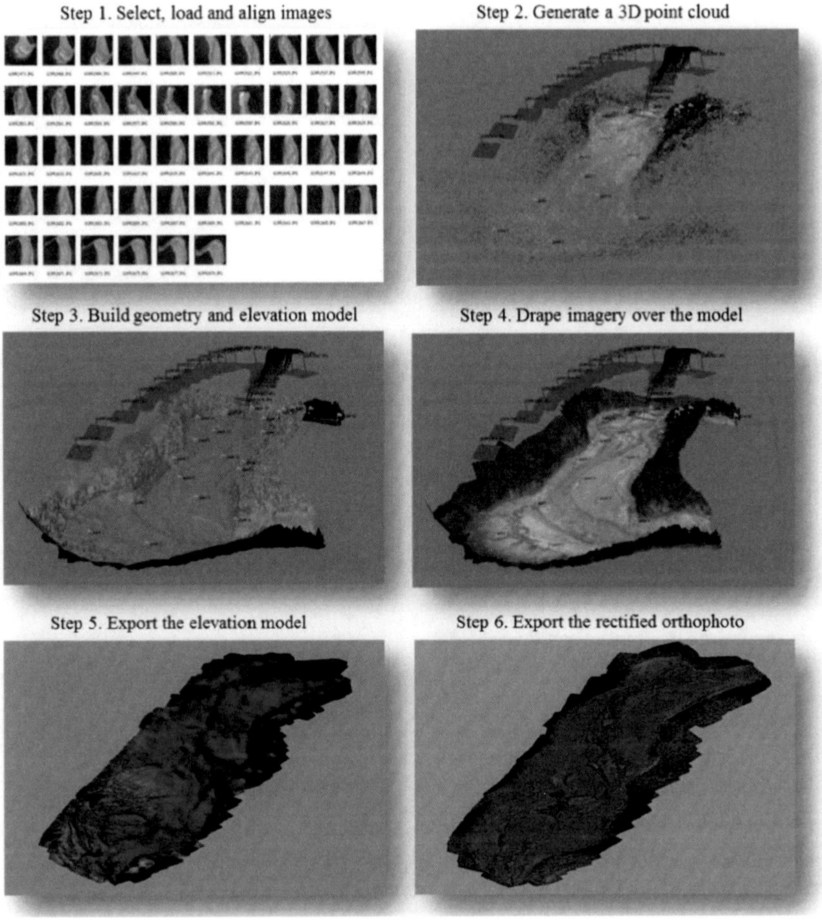


Figure 35. Images showing steps in a geometric correction and 3-dimensional reconstruction that use acquired images.

### *Emergent Sandbar Habitats*

The central Platte River valley in Nebraska is an internationally recognized staging area for migratory water birds of the Central Flyway; it supports more than half a million sandhill cranes and several million other waterfowl that migrate annually (Birds and Nature, 1998). Changes in water and land use, which have transformed the river channel and altered adjacent wet meadows and the structure of riparian habitats, have brought into question the sustainability of migratory and resident birds and other biota (fig. 36). A primary concern of land managers is determining how river flow influences the heights of sandbars used by endangered or threatened nesting birds (interior least tern and piping plover). However, owing to the dynamic nature of river flows, measuring sedimentation processes at fine temporal scales (days) and at high spatial resolution (centimeters) for long river reaches (kilometers) by using traditional methods is both challenging and costly.

Figure 36. Image of the central Platte River valley in Nebraska. Image taken at 200 feet above ground level by a Canon PowerShot SX230 on a T-Hawk.

To see if data could be acquired more cost effectively by a small unmanned aircraft, in summer 2013 a T-Hawk, equipped with a high-resolution digital camera (Canon PowerShot SX230), mapped the spatial extent and elevation of emergent sandbars along two reaches of the Platte River in Nebraska.

Both of these reaches, managed by the Platte River Recovery Program, are pilot sites for sediment augmentation to offset deficiencies in sediment supply caused by upstream water regulation. The USGS already monitors river morphology and sedimentation at these sites. Suspended-sediment transport is quantified by using ground surveying along transects, traditional sampling of bedload and suspended sediment, and newer technologies such as laser diffraction and acoustic backscatter. The establishment of ground control supports the generation of orthophotographs and digital elevation models of the emergent sandbars (figs. 37 and 38). These field sites are close to USGS stream-flow gaging stations, making water-surface- elevation data available. Conventional real-time kinematic surveys will be used to compare and demonstrate the accuracy and efficiency of using a small unmanned aircraft to map emergent sandbars.

Most of the imagery acquired during the mission was collected with a Canon SX230 camera onboard a T-Hawk, flown at perpendicular transects at 200 feet and 400 feet above ground level. This flight path effectively collected survey-grade data; horizontal and vertical accuracies were measured in centimeters. This degree of accuracy in both vertical and horizontal measurements lends itself to many types of hydrologic monitoring, and shows great promise for supporting hydrographic modeling and measurements applicable to sediment-movement and augmentation studies.

Figure 37. Three-dimensional model and orthophotograph derived from central Platte River valley images shown in figure 36.

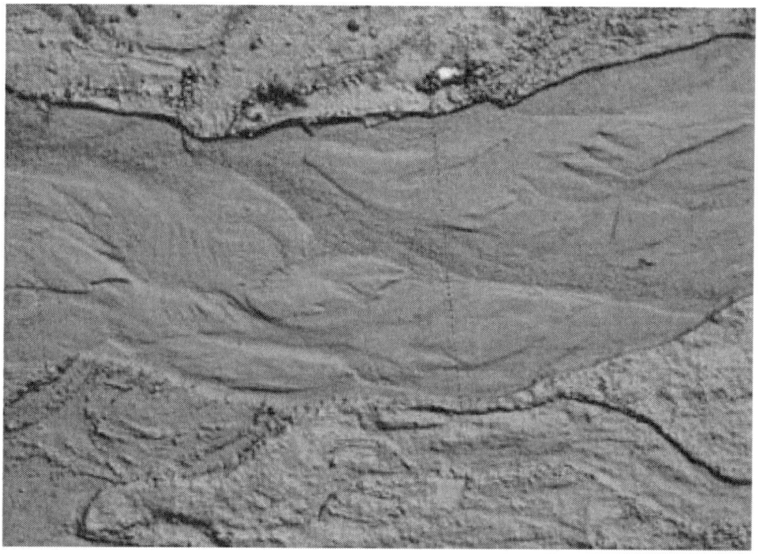

Figure 38. Elevation model derived from central Platte River valley images shown in figure 36.

## NEXT STEPS

Full integration of unmanned aircraft systems (UAS) into the Department of the Interior (DOI) will be based on proven operational procedures and benefits, such as increased safety and reduced cost, derived from completed proof-of-concept missions. A complete UAS program must also support a variety of operational models—from a single scientist whose frequent use of a small UAS (sUAS) requires it to be treated like other field equipment with direct purchase and operation, to those who have infrequent or unique UAS requirements. The more infrequent or unique needs could be fulfilled by working with the USGS National Unmanned Aircraft Systems Project Office (NUPO) or the nearest regional UAS hub, by use of a commercial UAS services contract, or in collaboration with other agencies or universities. Substantial progress has already been made towards integrating UAS into the DOI, from establishing a memorandum of understanding between the DOI and the FAA, to the successful completion of a multitude of proof-of-concept missions. But work is still needed to reach full integration. The USGS NUPO will continue to work towards full integration through the following actions:

- evaluate the next generation of UAS capabilities and potential commercial services;
- participate in domestic policy activities with Office of Aviation Services, Federal Aviation Administration (FAA), National Aeronautics and Space Administration (NASA), National Oceanic and Atmospheric Administration (NOAA), Department of State, and others;
- establish additional partnerships with DOI bureaus, other Federal agencies, and others to leverage missions and share technology;
- implement new geospatial datasets and products for use with UAS imagery; and
- address data storage and management needs for UAS acquired data.

Strategic actions are presented below. They are based on lessons learned during small unmanned aircraft flights, coordination with an international community of unmanned aircraft systems experts, and active interaction with professional societies, academia, and private industry. The actions are grouped into three time periods: near-term (next 12 months), longterm (next 3–5 years), and continuous.

## Near-Term Actions—Within the Next 12 Months

### *Transition to Commercial Small Unmanned Aircraft Systems*

The Raven A, vintage 2005, and T-Hawk, vintage 2008, were provided by the Department of Defense; they have been invaluable for evaluating the potential of small unmanned aircraft systems (sUAS) to support the numerous, complex, and varied Department of the Interior missions. However, both were designed for military applications, so they use military frequencies that require military spectrum approvals in addition to the National Telecommunications and Information Administration approval. They are also more expensive to use and maintain as compared with today's commercial sUAS, and have dated user interfaces, limited support for new sensors, and poor mission-support tools. Addressing the potential obsolescence of these systems accelerates the need to procure both commercial fixed-wing and vertical takeoff and landing sUAS for operational tests and evaluation.

*Recommendation:* Proceed with the bureau system requirements process currently underway and work with Office of Aviation Services (OAS) to establish streamlined procurement strategies for new sUAS (after evaluating potential systems for an "indefinite delivery/indefinite quantity" type of contract). Many potential systems are commercially available, support an international market and may have already operated in the National Airspace System.

*Action Assignment:* U.S. Geological Survey and OAS.

Several other actions are related to or required for the transition to new commercial systems.

### *Radio Spectrum*

Unmanned aircraft systems (UAS) require radio signals (spectrum) for flight control and to download the images being acquired by the sensor. The National Telecommunications and Information Administration (NTIA) manages the Federal government's use of spectrum, ensuring that domestic and international spectrum needs are met while making efficient use of this limited resource. The NTIA agency and Department of the Interior (DOI) spectrum offices must approve the spectrum used during each DOI UAS flight—a process that can currently take as much as 3 months. This lengthy review period hinders the use of UAS by DOI bureaus, especially if one considers the more "file and fly" certificate of authorization process outlined in the memorandum of understanding between the Federal Aviation Administration and the DOI.

*Recommendation:* In the near term, purchase new commercial systems that use public frequencies and don't require spectrum approval, which is how academic institutions operate. The USGS National Unmanned Aircraft Systems Project Office would coordinate with the USGS and DOI spectrum offices to establish hopping, encryption, and other system-hardening requirements to address any security issues with these frequencies. In the long term, within the next 3 to 5 years, move to spectrum dedicated to use by DOI sUAS.

*Action Assignment:* National Radio and Spectrum Program Management Office to identify long-term dedicated spectrum.

### *Airworthiness*

Certifying the airworthiness of unmanned aircraft systems (UAS) is a requirement for flying in the National Airspace System. Although the American Society for Testing and Materials (ASTM) has published a draft set of standards for small UAS (sUAS), the Federal Aviation Administration (FAA) has not established airworthiness criteria for UAS and instead leaves it up to the public operator to certify its own aircraft. The Office of Aviation

Services (OAS) is working diligently to establish an acceptable airworthiness process for the Department of the Interior. One option permits commercial vendors to have their UAS certified using the ASTM standards. Another option permits OAS to certify the UAS by using the ASTM standards or by contracting to a third party to test and verify the airworthiness of the commercial systems of interest to the DOI.

*Recommendation:* Develop a plan for certifying the airworthiness of new commercial systems, including a proposed schedule that will be accepted by both OAS and the FAA. *Action Assignment:* OAS and DOI aviation working groups.

## Long-Term Actions—In the Next 3 to 5 Years

### *Airworthiness and Training*

The Office of Aviation Services (OAS) currently trains (in Boise, Idaho) all operators of Department of the Interior small unmanned aircraft systems (sUAS); certifies the airworthiness of the systems; and approves any sUAS modifications (primarily sensor packages). With the large number of affordable sUAS currently available, this model will be difficult to maintain.

*Recommendation:* Transition to a model that requires airworthiness certification, and potentially operator training, to be provided by the OAS or by other government agencies (such as National Aeronautics and Space Administration, National Oceanic and Atmospheric Administration, Department of Defense), sUAS vendors, or accredited universities. OAS wouldl review and certify the airworthiness and training processes used by other organizations. It is also recommended that the "ground school" portion of the DOI UAS training program be converted to an on-line course.

*Action Assignment:* OAS

### *Large Unmanned Aircraft Systems*

Cost/benefit information makes it unlikely that the Department of the Interior (DOI) will have the budget flexibility to own or operate larger, more complex Class 3-5 unmanned aircraft systems (UAS) that can weigh as much as a 737 jetliner. But small UAS, the predominant UAS used by the USGS, do have limited sensor payloads and range; in certain situations, such as major flooding, hurricanes, wildfire, and arctic research, the DOI would still benefit from access to large long-duration UAS.

*Recommendation:* Ensure that the USGS Geospatial Product and Services Contracts are flexible enough to allow vendors to use any size UAS for acquiring and providing geospatial data. Modify existing agreements, or establish new agreements, with Federal agencies that may be able to support DOI missions in need of larger UAS platforms.

*Action Assignment:* USGS

***System Maintenance***

The USGS National Unmanned Aircraft Systems Project Office believes that a "you buy it, you break it, you fix it" model will become the standard practice for small unmanned aircraft systems (sUAS) maintenance. Although having the Office of Aviation Services (OAS) establish procurement contracts is appropriate, the concept of adding a "fleet" per-hour rate to use the sUAS might not. The usage fee concept works well for expensive and costly to maintain manned aircraft because it creates a pool of money to cover these operational costs, but inexpensive sUAS operated by bureau staff are more like computers, field vehicles, and other field equipment that only have a short life cycle.

*Recommendation:* The vendor can perform standard UAS maintenance. If a sUAS is beyond repair or obsolete, it should be surplused or discarded and replaced with a newer model. *Action Assignment:* USGS National Unmanned Aircraft Systems Project Office and OAS.

***Sensor Enhancement***

Many bureau missions could have overlapping requirements for unmanned aircraft systems (UAS) sensor payloads, which may lead to duplication of effort regarding identification, specification, testing, evaluation, and procurement. For example, on the basis of user demand, the Department of the Interior (DOI) might have a requirement to acquire enhanced thermal cameras in 2014, hyperspectral capabilities in 2015, and synthetic aperture radar in 2015.

*Recommendation:* Initiate an action for the DOI Remote Sensing Working Group to solicit for and prioritize UAS sensor requirements and acquire UAS platforms that are capable of interchangeable "plug and play" payloads.

*Action Assignment:* Office of Aviation Services, DOI Remote Sensing Working Group.

## Continuing Actions—External Coordination

The USGS exercises agreements with universities that are among the most respected in the United States for their knowledge of the technology of unmanned aircraft systems (UAS), use commercial UAS, and are willing to collaborate on UAS research and development of applications. The universities would maintain operational control of the systems and obtain the necessary Certificate of Authorization and airworthiness approvals accepted by the Federal Aviation Administration. This arrangement is very similar to one that the USGS and the University of Florida use today. The USGS is also working with National Oceanic and Atmospheric Administration, Department of Homeland Security, National Aeronautics and Space Administration, and Mesa County, Colorado, on possible collaborative missions.

*Recommendation:* Establish agreements with universities or use standing Cooperative Ecological Study Units partnerships

*Action Assignment:* USGS National Unmanned Aircraft Systems Project Office, Department of the Interior bureaus

## Rapid-Response Missions

The full Federal Aviation Administration (FAA) Certificate of Authorization process is complicated and can be very lengthy, which has hindered the Department of the Interior's (DOI) ability to perform and evaluate unmanned aircraft systems missions that require a rapid response (such as support for wildland fires or other natural hazard events). But now that a memorandum of agreement has been established between the DOI and the FAA, the DOI is more likely to consider UAS as a tool for supporting these activities. Although these rapid-response missions share some characteristics with the missions that have already been described, they also introduce additional complexity into logistics and operations.

### *Incident Response—Wildland Fires*

By definition, a wildland fire has an ample fuel supply and is difficult to reach owing to its remote location or rugged terrain. The outcomes of wildland fires differ widely, but in extreme cases they encompass vast expanses of land, destroy extensive wildlife and wildlife habitat, cause irreparable damage to wilderness areas, destroy built structures, and may kill anyone unlucky enough to be caught in their path. Though wildland fires may occur any time during the year, their intensity and frequency depend on the season. For example during high fire season in the Western States, middle to late summer, as many as 10 to 20 fires may be reported in one day.

Variable terrain, high winds, and complex support logistics are just a few factors that contribute to the high degree of danger associated with wildfire management. These same factors also complicate the data collection activities needed to supply the real-time information about fire location, perimeters, direction of spread, burn intensity, and smoke plumes that is essential (fig. 39).

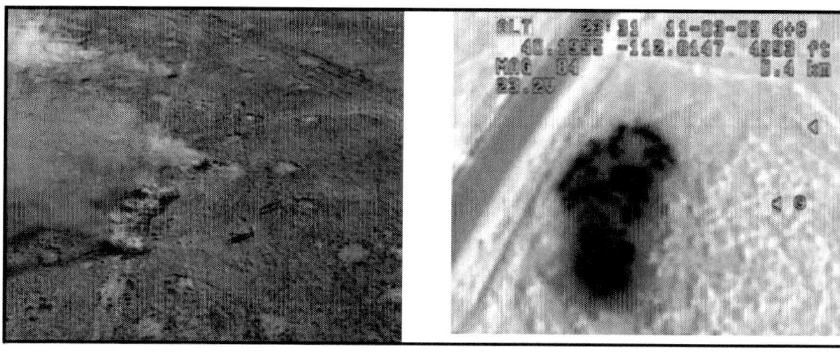

Figure 39. Electro-optical image (left) and infrared image (right) taken during a prescribed burn at Dugway Proving Ground, Utah.

During the day, when available and cleared for flight, observers in helicopters are commonly used to make wildland fire observations to establish data like fire location. But putting personnel in an aircraft at any time during fire operations is extremely risky owing to weather, low visibility due to smoke, and unpredictable fire behavior and intensity. Nightly infrared sensor operations can more safely and accurately establish fire locations but are not considered timely. Today, the best approach integrates satellite data, aerial imagery, and personnel positions obtained from a global positioning system (GPS); the resulting tactical data provide real-time situational awareness to the fire management's incident command center and the firefighters on the ground.

Satellite and other high altitude monitoring are used during very large wildland fires to provide a wider view. These more sophisticated systems employ GPS and aircraft-mounted infrared or high-resolution visible cameras to identify and target wildland fires. But satellite detection poses its own problems because it is prone to offset errors. For instance, moderate resolution imaging spectroradiometer and advanced very high resolution radiometer location data may be off by 2 to 3 kilometers (1 to 2 miles); geostationary operational environmental satellite location data may be off by as much as 12 kilometers (7.5 miles). Moreover, satellites in polar orbits commonly have only a brief observation time over a particular point, during which smoke or clouds may cover the location.

On the basis of these potential limitations, the Association for Unmanned Vehicle Systems International (AUVSI), a leading advocacy group for unmanned aircraft systems (UAS) for more than 30 years, has identified wildland fires and firefighting in remote lands as a prime candidate for civilian applications of UAS. In addition to tabletop exercises, AUVSI has been instrumental, along with the Department of the Interior (DOI) and the U.S. Forest Service, in conducting wildland fire UAS data collection exercises and simulations.

During a table top exercise in September 2010 at the U.S. Army's Dugway Proving Ground in Utah, AUVSI identified the following key objectives and associated findings:

***Objective 1.*** Determine how the availability of unmanned technologies may impact preparation activities, such as operational response planning and organizational coordination. Findings:

- The Interagency Wildland Fire Community is currently proactive in exploring areas for UAS integration into existing response systems, including other wildland fire management agencies.
- Business models for the employment of unmanned technologies in support of fire management services, such as mass broadcast text messaging (accomplished by the Israelis), already exist.
- Best practices from the military with regard to airspace deconfliction may be applied to operational firefighting.

***Objective 2.*** Identify the technical, regulatory, political, and organizational obstacles that may currently inhibit the use of unmanned systems in wildland firefighting applications. Findings:

- Technologies, such as Traffic Alert and Collision Avoidance System (TCAS), are available to provide effective Detect and Avoid (DAA) capability, which is a critical issue. Derivative DAA applications for addressing UAS integration to the NAS are actively under study.

- A substantial number of data sets are accessible, which can be used to build the safety case for the FAA to allow integration into the NAS.
- The FAA is currently collecting all information and data regarding UAS operations conducted under COAs.

***Objective 3.*** Develop insights and recommendations concerning the prospective operational utility, or shortcoming, of unmanned systems in firefighting applications and the transition to the general civil sector.

Findings:

- As a result of the live event at Dugway Proving Ground, participating systems successfully demonstrated the potential to provide persistent situational awareness.
- Many platforms are able to serve as a communications relay.
- The Table Top Exercise discussion revealed that there is a capability to provide continuous weather coverage.
- Small unmanned systems are transportable and equipped for rapid deployment.
- Versatility is achieved through the use of multiple sensors.
- The participating vehicles (UAS and Unmanned Ground Vehicles) proved to be capable of moving real-time data and information.
- Program maturity or the ability to capitalize on the lessons learned from the DOD.

***Objective 4:*** Examine the business case for operating unmanned technologies as wildfire management resources and the potential for other public safety applications. Findings:

- UAS could be of assistance to the Burned Area Rehabilitation Team.
- Support for revegetation (aerial seeding).
- Ability to perform fuel assessment.
- The majority of capabilities used to support the Fire Managers can also be applied to law enforcement and other public safety functions.
- Could provide persistent situational awareness.
- Supports a communications relay.
- Provides continuous weather coverage.
- Transportable and equipped for rapid deployment.
- Adaptable to the situation through the use of multiple sensors.

- Capable of moving real-time data and information.
- Additional applications include:
  - Wildlife monitoring and management
  - Soil erosion and postfire-debris-flow management
  - Road surveys

### *Natural Hazards—Volcanoes*

Volcanology is a compelling scientific discipline that develops deep-Earth science and a better understanding of the influence of magma on the shapes and forms of land and ocean-bed surfaces. Satellite imagery, such as Landsat imagery archived at the USGS Earth Resources Observation and Science Center, can serve as a reference with which to compare current observations of volcanic formations. As with monitoring any hazardous event, monitoring volcanic eruptions with manned aircraft introduces many potential safety hazards: gas emissions can interfere with internal combustion engine operations, gritty aerosols can clog turbo-machinery within high temperature aircraft propulsion systems, and abrupt release of volcanic magma can disrupt close-in aircraft operations by turbulence or reduced visibility. High-velocity rocks can be launched from primary or secondary vents and cause serious, potentially disabling damage to aircraft that are collecting critical preeruption information.

Sensors mounted on unmanned aircraft systems (UAS) can be safely used to acquire many types of data, such as surface temperature and composition of gases in the ejecta and in aerosols. UAS-based data collection would also avoid the personnel hazards associated with manned aircraft missions close to a dangerous, unpredictable, and potentially deadly natural hazard. Any indications of increased activity or near-surface activity can be pursued with higher-resolution imagery from multispectral and thermal imagers onboard UAS.

Close-in surveys including assays of active vents are possible without putting human life at risk by flying UAS through volcanic plumes, pyrocumulonimbus gas clouds, and ejecta.

### *Law Enforcement*

Several Department of the Interior (DOI) bureaus have law enforcement organizations that prevent, detect, and investigate crimes against resources on public lands. These crimes include theft—of mineral resources, of archeological and paleontological resources (or vandalizing them), and of timber, forest products, and native plants; dumping hazardous materials;

cultivation, manufacture, smuggling, and use of illegal drugs; illegal use of off-road vehicles; and wildland arson.

Aerial still and video photography support these efforts but are hampered by the size and remoteness of the public lands managed by the DOI and by increasing costs. In addition, satellite and other high-altitude platforms with fixed orbits can be limited by weather, and inaccessible or rough terrain can severely limit ground-based efforts.

High-resolution imagery from UAS could greatly improve the ability of DOI law-enforcement officials to collect more accurate real-time information in support of these activities. In fact, the Department of Homeland Security's Customs and Border Protection already utilize UAS as one of their strategic resources for routine mid-to-high-altitude data collection for border monitoring.

## Regulatory Challenges

Although the introduction of UAS into the National Airspace System (NAS) is just beginning, the commercial applications for these aircraft are as limitless as the ingenuity of their technical and entrepreneurial advocates. UAS automation and communication advances, such as intelligent flight-control systems, digitized communication protocols, and fully integrated flight planning will become an essential basis for advanced UAS interoperability in the NAS.

Since the Federal Aviation Administration (FAA) established the Unmanned Aircraft Systems Integration Office (UASIO) to integrate unmanned aircraft systems (UAS) safely and efficiently into the National Airspace System, it has been working on a number of issues. In order to safely integrate UAS into nonsegregated airspace, a robust detect and avoid (DAA) capability, a robust and secure command and control (C&C) data link capability, and separation assurance requirements for specified airspace need to be established.

To achieve these objectives, the UASIO and major UAS stakeholders are working closely with the UAS community to develop the minimum operational performance standards for DAA equipment and for the C&C data link.

Other issues are privacy, security, and environmental implications; developing acceptable design standards that consider aircraft size, performance, mode of control, intended operational environment, and mission criticality; and a regulatory process that is transparent to the public and provides an opportunity to understand and comment on proposed rules.

To date the FAA has released three documents—the Department of Transportation UAS Comprehensive Plan, the Final Privacy Requirements for the UAS Test Site Program, and the Integration of Civil Unmanned Aircraft Systems in the National Airspace System Roadmap. These documents and other FAA UAS related documents can be found at http://www.faa.gov/about/initiatives/uas/.

## REFERENCES

Association for Unmanned Vehicle Systems International, 2013, Economic report—Association for Unmanned Vehicle Systems International: accessed June 2013 at http://www.auvsi.org/econreport.

Birds and Nature, 1998, North American migration flyways: accessed July 29, 2011, at http://www.birdnature.com/flyways.html.

Bureau of Ocean Energy Management, 2013, Oil & gas energy program: accessed May 13, 2013, at http://www.boem.gov/Oil-and-Gas-Energy-Program/.

Bureau of Ocean Energy Management, Regulation and Enforcement, 2013, Home: accessed May 13, 2013, at http://www.boemre.gov/.

Bureau of Reclamation, 2011a, Bureau of Reclamation—About us: accessed July 29, 2011, at http://www.usbr.gov/main/about/.

Bureau of Reclamation, 2011b, Dam safety: accessed July 29, 2011, at http://www.usbr.gov/ssle/damsafety/.

Federal Aviation Administration, 2013, National policy N 8900.227, 43 p., accessed July 7, 2013, at http://www.faa.gov/documentLibrary/ media/Notice/N_8900.227.pdf.

Federal Aviation Administration, 2014, Aeronautical information manual, 726 p., accessed June 2014 at http://www.faa.gov/air_traffic/publications/media/AIM_Basic_4-03-14.pdf.

McCombs, Brady, November 23, 2010, Unmanned craft aid border effort, Arizona Daily Star: accessed July 29, 2011, at http://tucson.com/ news/blogs/border-boletin/border-bolet-npredator-bs-on-the-border/article_cb178a88-f687-11df-b113-001cc4c03286.html.

National Aeronautics and Space Administration, 2010, About MODIS: accessed December 7, 2010, at http://modis.gsfc.nasa.gov/about/.

National Commission on the BP Deepwater Horizon Oil Spill and Offshore Drilling, 2011, Deep Water—The Gulf Oil disaster and the future of offshore drilling: 398 p., accessed July 29, 2011, at http://www.gpo.gov/fdsys/pkg/GPO-OILCOMMISSION/pdf/GPOOILCOMMISSION.pdf.

National Park Service, 2010a, U.S. National Park Service—About us: accessed December 1, 2010, at http://www.nps.gov/aboutus/index.htm.

National Park Service, 2010b, National Park Service—NPS planning: accessed December 1, 2010, at http://planning.nps.gov/default.cfm.

Office of Aviation Services, 2013a, DOI operational procedures memorandum (OPM) no. 13-11: 6 p., accessed January 2014, at http://oas.doi.gov/library/opm/CY2013/OPM_13-11.pdf. Office of Aviation Services, 2013b, Office of Aviation Services mission statement: accessed March 18, 2013, at http://oas.doi.gov/mission.htm.

Office of the Secretary of Defense, 2009, Unmanned systems integrated roadmap 2009–2034: Washington, D.C., Government Printing Office, April 6, 2009, p. 96–97; available at http://www.dtic.mil/get-tr-doc/pdf?AD=ADA522247.

Office of Surface Mining Reclamation and Enforcement, 2008, Annual report 2008: 84 p., accessed November 16, 2010, at http://www.osmre.gov/resources/reports/2007-08.pdf. Office of Surface Mining Reclamation and Enforcement, 2010, About OSMRE—Who we are: accessed November 16, 2010, at http://www.osmre.gov/about.shtm.

Teal Group Corporation, 2013, Teal Group predicts worldwide UAV market will total $89 billion in its 2013 UAV market profile and forecast, June 17, 2013: accessed June 2013 at http://tealgroup.com/index.php/about-teal-group-corporation/press-releases/94-2013-uavpress-release.

U.S. Department of the Interior, 2011, Strategic plan for fiscal years 2014–2018: 56 p., accessed July 2011 at http://www.doi.gov/pmb/ppp/upload/DOI-Strategic-Plan-for-FY-2014-2018- POSTED-ON-WEBSITE-4.pdf.

U.S. Department of the Interior Indian Affairs, 2011, Indian Affairs—Who we are: accessed July 2011 at http://www.bia.gov/WhoWeAre/index.htm.

U.S. Fish and Wildlife Service, 2004, Migratory bird program strategic plan 2004–2014: accessed July 1, 2014, at http://www.fws.gov/migratorybirds/AboutUS.html.

# APPENDIX 1. ADDITIONAL READING

Department of Defense, 2011, Unmanned systems integrated roadmap FY2011–2036: Reference Number 11-S-3613, 108 p. Available at http://www.acq.osd.mil/sts/docs/Unmanned%20Systems%20Integrated%20Roadmap%20FY 2011-2036.pdf.

European RPAS Steering Group, 2013, Roadmap for the integration of civil remotely piloted aircraft systems into the European aviation system—Final report from the European RPAS Steering Group: 15 p. Available at http://ec.europa.eu/enterprise/sectors/aerospace/files/rpasroadmap_en.pdf.

Federal Aviation Administration, 2013, Integration of civil unmanned aircraft systems (UAS) in the National Airspace System (NAS) roadmap: Washington D.C., U.S. Department of Transportation, 71 p. Available at http://www.faa.gov/uas/media/uas_roadmap_2013.pdf.

International Civil Aviation Organization, 2011, Unmanned aircraft systems (UAS): Quebec, Canada, International Civil Aviation Organization (ICAO), 54 p., ICAO Cir 328; ISBN 978- 92-9231-751-5. Available at http://www.icao.int/Meetings/UAS/Documents/Circular%20328_en.pdf.

Joint Planning and Development Office, 2012, Next generation air transportation system— NextGen UAS research, development and demonstration roadmap version 1.0: 80 p. Available at http://fas.org/irp/program/collect/uas-nextgen.pdf.

U.S. Army, 2010, Eyes of the Army—U.S. Army unmanned aircraft systems roadmap 2010– 2035: 140 p. Available at http://www-rucker.army.mil/usaace/uas/US%20Army%20UAS%20RoadMap%202010%202035.pdf.

# APPENDIX 2. ABBREVIATIONS USED IN THIS REPORT

| | |
|---|---|
| AGL | above ground level |
| ASTM | American Society for Testing and Materials |
| AUVSI | Association for Unmanned Vehicle Systems International |
| BIA | Bureau of Indian Affairs |
| BLM | Bureau of Land Management |
| BOEM | Bureau of Ocean Energy Management |
| BSEE | Bureau of Safety and Environmental Enforcement |
| C&C | command and control |
| CBP | Customs and Border Protection |

| | |
|---|---|
| CDPA | California Desert Protection Act |
| COA | Certificate of Authorization |
| CPW | Colorado Parks and Wildlife |
| DAA | detect and avoid |
| DOD | Department of Defense |
| DOI | Department of the Interior |
| EO | electro-optical |
| FAA | Federal Aviation Administration |
| GPS | global positioning system |
| GPU | graphics processing unit |
| MSL | mean sea level |
| NAS | National Airspace System |
| NASA | National Aeronautics and Space Administration |
| NOAA | National Oceanic and Atmospheric Administration |
| NTIA | National Telecommunications and Information Administration |
| NUPO | National Unmanned Aircraft Systems Project Office (USGS) |
| NWR | National Wildlife Refuge |
| OAS | Office of Aviation Services |
| OSM | Office of Surface Mining Reclamation and Enforcement |
| sUAS | small unmanned aircraft systems |
| UAS | unmanned aircraft systems |
| UASIO | Unmanned Aircraft Systems Integration Office |
| USFWS | U.S. Fish and Wildlife Service |
| USGS | U.S. Geological Survey |

## APPENDIX 3. COSTS OF USING AIRCRAFT

As part of an article on Homeland Security's Unmanned Aerial Program (McCombs, 2010), examples of per-hour costs for a variety of aircraft were provided from the U.S. Customs and Border Protection Air and Marine group. The per-hour costs for the manned aircraft include fuel, airframe, engine, avionics, other equipment, support, and services; for the Predator B unmanned aircraft systems (UAS) the costs include satellite time, maintenance, and pilot training (table A3-1).

**Table A3–1. Cost per hour of aerial platforms from the Customs and Border Protection Office [UAS, unmanned aircraft system]**

| Cost per hour (dollars) | Platform | | Type | |
|---|---|---|---|---|
| 7,034 5,245 5,233<br>3,994 3,744<br>3,476 | P3<br>King Air C12<br>Blackhawk<br>King Air<br>Agusta Westland<br>AW139<br>UH1H (Huey) | Aircraft<br>Aircraft<br>Aircraft | <br><br><br><br>Helicopter<br>Helicopter<br>Helicopter | |
| 3,395 | Cessna Citation | Aircraft | | |
| 3,234 | UAS (Predator B) | | | UAS |
| 2,781 | Pilatus PC-12 | Aircraft | | |
| 2,568 | Dash 8 | Aircraft | | |
| 1,932 | Aerospatiale Astar | | Helicopter | |
| 1,841 | MD-600N | | Helicopter | |
| 1,727 | Cessna 210 | Aircraft | | |
| 1,617 | American Eurocopter EC-120 | | Helicopter | |
| 1,441 | Hughes OH6 | | Helicopter | |
| 1,141 | Piper PA-18 | Aircraft | | |

Source: McCombs, Brady, November 23, 2010, Unmanned craft aid border effort, Arizona Daily Star: accessed July 29, 2011, at http://tucson.com/news/blogs/border-boletin/border-bolet-n-predator-bs-on-theborder/article_cb178a88-f687-11df-b113-001cc4c03286.html.

# APPENDIX 4. UNITED STATES AIRSPACE

In the United States, National Airspace System "Controlled" airspace consists of Class A, B, C, D, and E; "Uncontrolled" airspace consists of Class G (fig. A4–1).

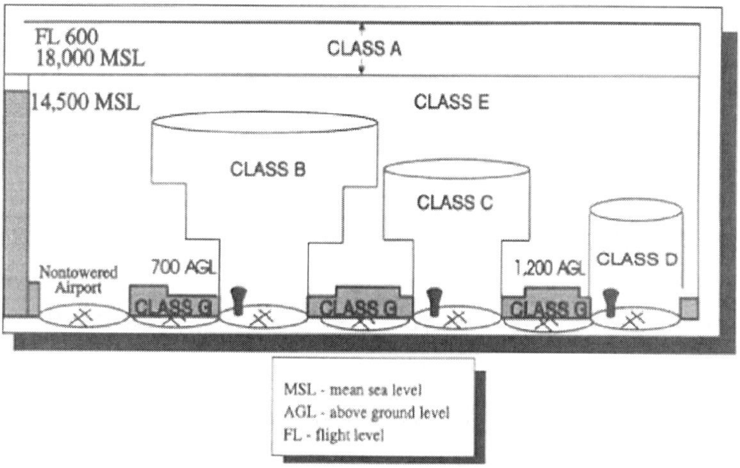


Figure A4-1. Airspace classifications in the National Airspace System (FAA).

## Class A Airspace

Class A airspace extends from 18,000 feet above mean sea level (MSL) to approximately 60,000 feet MSL including the airspace overlying the waters within 12 nautical miles of the coast of the 48 contiguous States and Alaska. Unless otherwise authorized, all flights must operate under Instrument Flight Rules (IFR).

## Class B Airspace

Class B airspace is defined around key airport traffic areas, usually airspace surrounding the busiest airports according to the number of IFR operations and passengers served.

## Class C Airspace

Class C airspace is defined around airports of moderate importance that have an operational control tower, are services by a radar approach control, and have a certain number of IFR operations.

## Class D Airspace

Class D airspace normally extends from the surface to 2,500 feet AGL surrounding airports that have an operational control tower. This class reverts to Class E or G during hours when the tower is closed or other special conditions.

## Class E Airspace

Class E airspace is any "controlled" airspace that is not Class A, B, C, or D, which applies to most of the airspace in the United States. In most areas it extends from 1,200 feet AGL up to but not including 18,000 feet MSL, the lower limit of Class A airspace.

## Class G Airspace

Class G airspace includes all airspace below flight level 600 not otherwise classified as controlled. This typically includes the airspace very near the ground (1200 feet or less), beneath Class E airspace, and in areas completely uncontrolled.

See Chapter 3 of the FAA Aeronautical Information Manual for additional information (Federal Aviation Administration, 2014).

In: U.S. Geological Survey Roadmap …
Editor: Kathryn Miller
ISBN: 978-1-63483-566-4


*Chapter 2*

# U.S. GEOLOGICAL SURVEY UNMANNED AIRCRAFT SYSTEMS PROJECT OFFICE LESSONS LEARNED AND OPPORTUNITIES*

*Michael E. Hutt*

## ABSTRACT

The U.S. Geological Survey Unmanned Aircraft Systems (UAS) project office is leading the implementation of UAS technology in anticipation of transforming the research methods and management techniques employed across the Department of the Interior. UAS technology is being made available to monitor environmental conditions, analyze the impacts of climate change, respond to natural hazards, understand landscape change rates and consequences, conduct wildlife inventories and support related land management missions. USGS is teaming with the Department of the Interior Aviation Management Directorate (AMD) to lead the safe and cost effective adoption of UAS technology by the Department of the Interior Agencies and USGS scientists. Recently, DOI-USGS sponsored our second UAS training session. Each student received instruction in basic and advanced flight skills, airspace management, aviation safety, emergency procedures, crew coordination, and mission planning.

* This is an edited, reformatted and augmented version of a document issued in 2010.

## INTRODUCTION

The mission of the U.S. Department of the Interior is to protect America's natural resources and heritage, honor our cultures and tribal communities, and supply the energy to power our future. The U.S. Department of the Interior uses sound science to manage and sustain America's lands, water, wildlife, and energy resources, honors our nation's responsibilities to tribal nations, and advocates for America's island communities. As keepers of our nation's legacy, we manage the resources in our care to benefit Americans now and in the future. Interior employees are developing and implementing the cutting-edge science and expert management techniques that make this possible. We are also taking actions to help America transform the challenges it faces today into opportunities for tomorrow. The Department employs 70,000 people, including expert scientists and resource-management professionals, in eight technical bureaus:

- Bureau of Indian Affairs
- Bureau of Land Management
- Bureau of Ocean Energy Management, Regulation and Enforcement
- Bureau of Reclamation
- National Park Service
- Office of Surface Mining, Reclamation and Enforcement
- U.S. Fish and Wildlife Service
- U.S. Geological Survey

Since Congress created the Department of the Interior in 1849, it has become the steward for managing over 500 million acres of surface land, or about one-fifth of the land in the United States. The Department has responsibility for managing a variety of water and underwater resources. The Bureau of Reclamation manages nearly 500 dams and 350 reservoirs that deliver irrigation water to one of every five western farmers and provides water to over 30 million people. The Bureau of Ocean Energy Management, Regulation and Enforcement has jurisdiction over approximately 1.76 billion acres of the Outer Continental Shelf, on which it manages approximately 9,000 active oil and gas leases on 44 million acres. The Department hosts over 460 million visitors to nearly 5000 recreational sites including wildlife refuges, parks, monuments, seashore sites, battlefields and other cultural and recreational sites on a yearly basis. The energy projects on federally managed lands and offshore areas provide about 30 percent of the nation's energy production. The Department's law enforcement workforce is one of the largest in the Federal Government.[1]

The U.S. Geological Survey (USGS) serves the Nation as an independent fact-finding agency that collects, monitors, analyzes, and provides scientific understanding about natural resource conditions, issues, and problems. The value of the USGS to the Nation rests on its ability to carry out studies on a national scale and to sustain long-term monitoring and assessment of natural resources. Because it has no regulatory or management mandate, the USGS provides impartial science that serves the needs of our changing world. The diversity of scientific expertise enables the USGS to carry out large-scale, multi-disciplinary investigations that build the base of knowledge about the Earth. In turn, decision makers at all levels of government-and citizens in all walks of life- have the information tools they need to address pressing societal issues. As an unbiased, multi-disciplinary science organization that focuses on biology, geography, geology, geospatial information, and water, we are dedicated to the timely, relevant, and impartial study of the landscape, our natural resources, and natural hazards.[2] Beginning in Fiscal Year 2005, the U.S. Geological Survey, Land Remote Sensing Program (LRS) sponsored the Rocky Mountain Geographic Science Center to conduct an Emerging Technology Investigation on Unmanned Aircraft Systems (UAS). The goals of the Emerging Technology Investigation Projects are to monitor promising technologies and determine their potential to support the USGS and DOI mission. Often the investigations involve the creation of Cooperative Research and Development Agreement (CRADA) or similar agreements with academia, private industry or governmental entities. The LRS program has supported investigations into sensor web, hyperspectral imaging, data compression and similar geospatial technologies. The findings of the unmanned aircraft investigation included:

- Military use of UAS technology was increasing at an exponential rate,
- U.S. investments in UAS technology for 2009-2013 were projected to be between \$25 - \$30 Billion,
- The Department of Defense, NOAA, NASA, DHS, and the FAA had or were in the process of establishing UAS Program Offices,
- 60 Universities in the U.S. were offering UAS related programs,
- There were over 150 vendors of UAS technology,

- A wide variety of platforms (aircraft) and sensor packages were available,
- Substantial momentum existed in industry, academia and government to develop civil applications of UAS technology,
- UAS technology could contribute to the safety of DOI aircraft crews & field staff,
- UAS provided large area coverage (compared to ground observations),
- UAS provided long dwell times over areas of interest,
- Readily available technology- minimum DOI/USGS research and development cost,
- Ability to operate in hostile environments (smoke, ash, active volcano, etc.),
- Relatively low costs to operate (sUAS) compared to manned aircraft,
- Provided a archive of images for future use and analysis,
- Puts technology in the hands of the scientists,
- New observations = new science = more informed decisions

These finding led to the USGS announcing the creation of a UAS Project Office on May 8, 2008. Many of the goals and objectives described in the announcement of the office still serve as guidance and direction for our current activities:

*In dangerous and remote areas, such as polar regions, volcanic islands, and expansive deserts, remote-controlled unmanned aircraft can provide more detailed, more timely data about the status of natural resources and environmental conditions than would be feasible by any other means. That is why the U.S. Geological Survey - long known as an authoritative source of aerial photography and satellite-based imagery - today announced that it is establishing a new program for earth observation using Unmanned Aircraft Systems (UAS).*

> "This exciting approach to earth observation gives scientists a way to look longer, closer, and more frequently at some of the most remote areas of the Earth, places that were previously too dangerous or too expensive to monitor in detail," said Barbara J. Ryan, (former) USGS Associate Director for Geography. "The flexibility and relative low cost of unmanned aircraft systems will enhance our ability to track long-term landscape change. In addition, we can quickly assess landscape altering

events, such as wildfires or volcanoes, in areas with challenging logistics."

*In many cases, UAS technology is simply the most cost effective way to gather earth observation data for a wide variety of applications: managing federal lands; investigating climate change; mapping and charting; conducting environmental risk assessments; responding to and recovering from natural and human-induced disasters.*

*Even in less remote areas, manned aircraft flights may not be feasible at times due to long flight durations, hazardous weather conditions, and associated operations cost. Satellite-based observations can be hindered by coarse image resolution, limited sensor capabilities, and repeat orbiting cycles of days or weeks. The use of UAS technology allows flexibility in delivering timely data. Furthermore, data collection by UAS can be specifically tailored to the required resolution and radiometric parameters of individual investigations.*

*An important focus of the USGS UAS Program, a component of the USGS Land Remote Sensing Program, is to leverage the commitment that the defense and intelligence communities have made in supporting UAS research. Working in partnership with many other Federal agencies, academia, and industry groups, the USGS will promote UAS technology for civil, domestic applications. Offices for the new program will be located at the USGS facility in Lakewood, Colo.*[3]

## Opportunities

A focus of USGS Unmanned Aircraft Project Office is to maintain and enhance our ability to leverage the substantial commitment private industry, academia and the Defense Communities have made supporting advanced UAS research, technology and applications development. A project goal is to encourage industry and academic colleagues that have previously developed UAS technologies for national defense applications to support the evolution of a dual use capability that will benefit the DOI and USGS missions. The USGS UAS Project Office, is actively working with mission partners such as the National Oceanic and Atmospheric Administration (NOAA), National Aeronautics and Space Administration (NASA), United States Northern Command (NORTHCOM), Department of Homeland Security (DHS), United States Army, United States Forest Service (USFS), academia (notably- New

Mexico State University, University of Alaska- Fairbanks, University of Colorado, Virginia Tech. University, North Dakota State University, John Hopkins University, University of Florida, and the U.S Air Force Academy).

The Department of the Interior, National Business Center, Aviation Management Directorate (AMD) is a valued partner of the U.S. Geological Survey Unmanned Aircraft Project Office. While the USGS UAS Project Office worries issues such as sensor research and collecting project requirements, the AMD is working tirelessly to ensure that the Department establishes, maintains and enforces safe aircraft operations. AMD provides the Department with several UAS related services including: Aviation Safety Programs, Aircraft Management Services, and User/Operator Training. After an exhaustive review of existing standards and participating in the sUAS operator training session, AMD recently issued guidance- *DOI AVIATION MANAGEMENT OPERATIONAL PROCEDURES (OPM) MEMORANDUM NO. 09-11, DOI Use of Unmanned Aircraft Systems (UAS).* The OPM establishes UAS operational procedures, aircraft maintenance criteria, aircrew qualifications and maintenance personnel credentials for use across the Department of the Interior. A link to the OPM is available at: http://rmgsc.cr.usgs.gov/UAS/[4]

Other AMD program responsibilities and goals include increasing the efficiency and promoting the economical operation of aircraft, including UAS, for the Department of the Interior. AMD provides the Department with expert technical assistance in specialized and emerging aviation technology areas such as UAS integration into the National Air Space. The USGS UAS Project Office is working closely with AMD on determining the appropriate avenue to obtain UAS capabilities and services including: procuring unmanned aircraft, establishing commercial UAS aviation services contracts, and coordination of utilization of UAS assets with other governmental entities.

UAS capabilities have demonstrated the potential to effectively fill the current observation gaps that are critical to gaining a better understanding of the complexities and scientific knowledge related to climate change research, water resources forecasting, ecosystem monitoring and management and natural hazards. These information gaps in our observations frequently exist over the remote, scarcely populated and often volatile lands managed by the Department of the Interior (North Slope of Alaska, Volcanic Islands, Everglades, the vast, desolate areas found in the intermountain West) and other remote reaches of the Earth. Manned aircraft flights are problematic in these areas due to long flight durations, unpredictable weather conditions and associated operations cost. Satellite based observations are hindered by course resolutions, limited sensor capabilities, weather conditions and acquisition cycles that can often be several days or weeks. The use of UAS technology allows flexibility in delivering timely data, for longer durations (persistent staring), tailored to the required resolution and radiometric parameters and expands the ability to obtain remotely sensed data to inventory and monitor dynamic landscape altering events and conduct impact analysis over previously logistically challenging areas. The use of UAS technology will enhance our ability to provide unbiased scientific information to better enable decision makers to make informed decisions. Unmanned Aircraft Systems (UAS) will transform the methods and techniques employed across the Department of the Interior (DOI) and the United States Geological Survey (USGS) in support of our mission to serve the Nation by providing reliable scientific information to describe and understand the Earth; minimize loss of life and property from natural disasters; manage water, biological, energy, and mineral resources; and enhance and protect our quality of life. Cost effective UAS technology is currently available to support a wide variety of applications including: managing federal lands; monitoring environmental conditions and natural resources use; analyzing dynamic earth processes to support global and climate change investigations; generating mapping, charting, and geodesy products; conducting environmental risk assessments; and preventing, preparing for, responding to, and recovering from natural and human-induced disasters. Collaboration across government, commercial entities, and academia are essential for the U.S. Geological Survey Unmanned Aircraft Project Office to efficiently and effectively evaluate and develop UAS capabilities that support and/or serve as:

- Augmentation of Satellite and Aircraft observations
- Verification and validation of remotely sensed observations
- Strategic Awareness in support of Incident Command Operations
- Wildland Fire Tactical Support
- DOI Law Enforcement (crowd control, Border Patrol, counter drug operations)
- Monitoring DOI assets (archeological sites, areas of critical habitat, critical infrastructure)
- Mapping ( Digital Elevation Models, Image Collection)
- Land Management (monitoring rangeland and forest health conditions)
- Climate Change (monitoring carbon exchange)

- Logistical Support (instrumentation of seismic/ GPS sensors; material resupply)
- Communications and Broadcast Services (enhanced radio communications)
- Search and Rescue operations
- Environmental Research
- Wildlife Inventories
- Air quality measurements (volcanic gas and ash emissions, wildfire smoke plume gasses)

Recently, the U.S. Geological Survey obtained several sUAS Raven Systems. The systems have provided us with the ability to compare our current baseline of capabilities (i.e. flight duration, sensor packages) against our mission requirements. This analysis will be critical as we move forward in our use of UAS technology. The systems have also presented us with the opportunity to work with AMD and the FAA on the definition of operational (crew training, airworthiness criteria, etc.) requirements and Certificate of Authorization (COA) process. The AeroVironment produced Raven is a small unmanned aerial systems (sUAS) that are primarily designed as a reconnaissance and surveillance tool. The system transmits near real time airborne video images, compass headings and location information to a ground control unit (GCU) and remote video terminal (RVT). The Raven aircraft is hand launched without the need of special equipment or landing areas. The system employs a self-stabilizing aircraft configuration with stability augmentation avionics and provides ease of control and steady video imagery. The optics package includes an electro-optical (EO) color camera nose payload (side and forward looking) on the same payload and two infrared (IR) thermal nose payloads (a side look payload or a forward look payload). The Raven system is typically operated by a two person team consisting of a Pilot and Mission Controller. The systems are proving invaluable in promoting and developing sUAS technology for civil and domestic applications.

The USGS UAS Project Office, plans to utilize this technology to support various scientific studies which include; analysis of wildfire and environmental impacts; monitoring land change; examining the affects of the mountain pine park beetle of the western states; tracking wildlife; completing ecosystem surveys; improving natural hazard forecasting and mitigation; and to aid and assist in GIS mapping & remote sensing projects within the department. The USGS and AMD have sponsored sUAS- Raven training sessions at Dugway Proving Grounds and Ft. Bliss. The two week courses, mirrored after training developed for the U.S. Army, have been attended by students representing USGS, NPS, BLM, AMD, FAA and USFS. Information on the course is available at: http://oneinterior.doi.net/oneinterior/News/nbc-usgs-unmanned-aircraft-systemtraining.cfm

## Challenges

The Project Office is representing USGS interest and supporting initiatives designed to implement UAS capabilities as a viable, cost effective and regulated alternative to existing land imaging resources. A partial listing barriers and areas that need further investigation:

- Operations Costs (compared to manned aircraft)
- Lack of commercial (contract) availability of UAS service providers
- Sensor familiarity and data exploitation tool availability
- Frequency availability to support Civil/ Commercial Operations
- Operator Training Certification Programs
- Data Archive- Dissemination policies and capabilities
- Concern with proliferation of "one of a kind systems"
- Lack of airspace available to maintain operator training and currency requirements
- Lack of airworthiness standards, especially for sUAS
- Few opportunities to document "real missions"
- Definition of "accident or incident"
- Liability for Civil Operations
- FAA National Air Space Regulations[5]

The most notable of the current restrictions and concerns is directly related to the Federal Aviation Administration Certification of Authorization process. The release and adoption of formal rules and regulations governing UAS operations in the National Airspace is eagerly anticipated by the USGS and our partners. It is hoped that the proposed rules recognize the high potential value of operating "program of record" sUAS as public aircraft, by trained operators and the relatively small risk associated with operating sUAS over the lands managed by the Department of the Interior. The majority of our remaining concerns will be rectified as systems and applications evolve and as we gain experience with this rapidly evolving technology.[6]

## CONCLUSION

The USGS UAS Project Office is dedicated to leading and coordinating the USGS efforts to promote and develop UAS technology for civil, domestic applications while not compromising risk to the public's safety. A key task, currently underway, is to develop a report that will serve as a roadmap for the development of these applications by the Department of the Interior and USGS. The document will be updated annually and used to evaluate the feasibility of future research, application developments, and mission support. In addition, it will help influence budget requests, plans and allocations. The report will:

- Provide an overview of the present state of the platforms and sensor capabilities,
- Provide a foundation for development of a comprehensive DOI/USGS UAS roadmap
- Determine and document technologies necessary to support future missions
- Document requirements for UAS and related technologies
- Determine and document future civil missions for UAS based on user-defined needs

The ultimate goal of the USGS Unmanned System Project Office is to support informed decision making that will directly benefit the Nation by creating the opportunity for USGS and our partners gain access to an increased level of persistent monitoring of earth surface processes (forest health conditions, monitoring wildfires, earthquake zones, invasive species, etc.) in areas that have been difficult or impossible to obtain information before. Much like the use of internet technology and global positioning systems, unmanned aircraft systems have the potential of enabling us to be better stewards of the land by:

- improving natural hazard forecasting and the analysis of the impacts;
- improving our understanding of climate change to better plan for the likely impacts;
- developing precipitation and evaporation forecasting to better manage water resources;

- monitoring Arctic ice change and its impacts on ecosystems, coasts, and transportation;
- increase safety and effectiveness of wildland fire management;
- broaden our abilities to monitor and enforce land use regulations;
- better understand and protect ecosystems

Harry Kieling, AMD Alaska Regional Director and student in the first Department of the Interior Unmanned Aircraft System class said:

> "A big segment of the future of aviation will include Unmanned Vehicles. They will not replace manned aviation, but a lot of DO! missions will be better and more safely performed by UAS. This Raven training is a big first step for DO! toward that goal."

## ACKNOWLEDGMENTS

Dr. Bruce Quirk- USGS, Mr. David Johncox- USGS, Mr. Dallas Keller-Aerodyne, Colonel Gregory Gonzales- U.S Army, Mr. Stan Wilds- Parallel Inc., Ms Robbie Hood- NOAA, Ms Brenda MulacNASA, Mr. Steven Hottman- NMSU, Mr. Harry Kieling- DOI/AMD and their staffs for their invaluable support to the U.S. Geological Survey UAS Project Office.

## APPENDIX

Below is a cross sample of requests related to potential applications of UAS technology provided by Department of the Interior employees. The list is provided to illustrate the ground swell of interest (including scientific, managerial and operations staffs) across the Department related to this emerging technology. To further illustrate the growing interest in UAS technology, over 100 Department of Interior employees have expressed an interest in being trained as sUAS operators.

> "Our scientists have expressed interest in the capabilities of this system to collect low-altitude aerial imagery used for bird and habitat studies."
>
> "although the FAA rules and regulations are covered in the course, it would be useful to have some general idea what these are to be able

to pass along to prospective students. Water Centers in parts of the country where use of the sUAS may be problematic would be less likely to take advantage of this opportunity."

"I've been saying for years that USGS needed a capability like this one appears to have, for stream geomorphology and riparian ecology work, detailed high-res imagery of surface conditions in small watersheds, and for flood or ice-jam monitoring. I see good use of this platform for a variety of uses... flood monitoring, spill detection and monitoring in rivers, coastal spill detection and monitoring."

"Our team supports post-earthquake scientific investigations following large earthquakes in the US and, on occasion, internationally. These investigations require rapid assessment of the extent of surface faulting, earthquake- induced landslides and liquefaction, and the damage area, and effective deployment of field teams to record critical scientific observations before many of the subtle features are destroyed by natural or man-made processes. Use of small unmanned aircraft by our field teams may have the potential to complement existing reconnaissance methods that we currently employ, by possibly reducing the amount of time-consuming ground-level reconnaissance that typically occurs during these investigations, or to access areas that otherwise would be impossible or unsafe for our field crews and scientists to enter."

"To combat multinational drug traffic organizations engaged in cultivation of large marijuana plantations on public lands. I am interested in UAV/UAS programs as a way to conduct pre-raid reconnaissance of routes and locations, assess damage to natural resources as a result of cultivation activities and aid in the collection of geospatial data on impacted lands."

"I'm leading wetland restoration programs in San Francisco Bay and on Puget Sound. We've been experimenting with aerial photography to document wetland change."

"One of our technical interests is the interaction of surface water and groundwater. In places where groundwater flows into surface waters, and a temperature difference exists between the waters, such locations can be identified by temperature. One way in which this is done is with fiber-optic cable. The combination of camera and thermal imager of the sUAS would provide another way to evaluate the location of groundwater flow to surface waters, as well as providing clues about where deployment of continuous recording devices such as the fiber optic cable, or other measurement techniques would be most effective. The surface waters of interest include streams, rivers, lakes, wetlands, estuaries and coastlines."

"I would like to conduct video or photo surveys for pelagic birds on the Great Lakes using some sort of line transect sampling scheme. Altitude of flight would depend on the optical equipment. Pelagic bird surveys on Great Lakes can be accomplished with a small unmanned aircraft vehicle."

"I have discussed the potential applications of aerial surveillance and we feel that this tool could improve the quality of the science we provide to our cooperators and the public. North Dakota is prone to flood events due to limited topographic relief (especially in the Red River Valley) and the rapid release of water from a melting snow pack in the spring. We had a historic peak with widespread flooding this past spring in both the Western and Eastern sides of the state simultaneously that threatened North Dakota's largest population centers. Aerial surveillance wouldn't mitigate flooding but it would allow the USGS to document and better respond in real-time to ice jams, overland flooding, and levee failures. This spring saw all of these events and the use of an aerial surveillance system would have allowed the USGS to provide better science to our sister agencies and cooperators as they attempted to mitigate flood damage. My time with the ND USGS has enabled me to lead on the use and application of developing technologies in the hydrologic field. The use of real-time aerial surveillance to capture hydrologic events has the potential to improve our science and I would like to be a part of this developing technology."

"If this resource becomes available, I may be able to use it in conjunction with a small hyper spectral sensor to collect imagery over Fossil Butte National Monument as part of a project to collect spectral signature information for a couple of at-risk plant species. There is concern that these species may be threatened by the growing elk population "

"I would be interested in using this for glacier monitoring using LIDAR and digital photography."

"Our uses of UAS would include: Bird population counts in wilderness or non-accessible terrain, fire ecology mapping, swan signature, population and nesting sites"

"Global Change - Habitat and marsh detection / thermal ground water discharge (brine / geochemistry) detection."

"We would like to determine the chemical composition of materials in smoke plumes, especially fires that occur in the wildland- urban interface."

"We are interested in evaluating the potential of a UAS to monitor wildland fire and disseminate the information to the field crews (using PDA technology), incident command center, Geographic Area Coordination Centers and National Interagency Fire Center."

"I am very interested in the potential for thermal imaging of streams and wetlands to assess groundwater influxes and thermal dynamics. I am also interested in low-cost documentation of canopy changes from close range. We would be able to deploy the instrument locally in VA over public (conservation reserve and/or National Park) lands to gain experience. I could then see us deploying the system in the Everglades and some other areas of the southeast where surface/subsurface water interactions are of great interest."

"Potential applications of interest would be post-fire severity assessment and high spatial/temporal resolution monitoring; characterization and monitoring of other disturbances at high spatial/temporal resolution, incl. for ephemeral targets/phenomena (e.g., pests/pathogens); field data augmentation and plot supplementation."

"We are interested in using high resolution remote sensing to map changes in vegetation type, cover and stress as well as temporal changes and spatial distribution of water quality. Our group specializes in measuring water color in situ to study biogeochemical aspects of water quality parameters (ancillary and dissolved and particulate organics and inorganics). We would like to evaluate the utility of using the RAVENS highly resolute, visible + IR (and spectral if possible) images to extend the reach of our in situ water quality (limited to point) measurements. Specifically, we hope to use the RAVEN to map the accretion of carbon in experimental farmed wetlands in the Sacramento Delta and to map the extent and evolution of riparian zones of a model agricultural watershed. "

"Wild horse population estimates of the herds on BLM lands near Cody, WY"

"We want to integrate collected imagery and data into existing GIS/GoogleMaps databases (with the help of USGS GIS programmers and geographers) so that the information will be easily shared with both fellow USGS colleagues and regional cooperators. We have many more applications in mind, too many to list here right now, but hopefully we can discuss them soon"

"The UAS would be used for river corridor flights along the upper Colorado River (Upstream of Williams Fork) and Fraser River in Grand County. Fly the river to track habitat conditions and potential changes over the years as more water is removed and transported to the Front Range. This project would be repeated on some time interval (every 3 to 5 years)."

"The UAS would be used to monitor two-dimensional river hydraulics and habitat mapping on the Cache La Poudre River near Fort Collins. USGS has a long-standing study site in the Poudre River Canyon, just below the State Fish Hatchery between Kinikinik and Idylwilde, CO. The additional imagery would add to our project looking into hydraulic model sensitivity to various data sources and densities which impact the final habitat maps we create for resources managers. We have been discussing creating a second study site along the river near Greeley in order to begin looking at how the stream bed morphology plays a part in our model sensitivity. Basically, the UAS would help compare a steep gravel bed channel within the Poudre River Canyon to the flat sand bed channel on the Plains."

"The UAS would support moose inventory flights using the infrared camera. It is difficult to locate moose via manned aerial flights due to the trees and the dark color of the moose. It may be possible to pick moose up with the infrared camera. Possible locations for these flights would be the Williams Fork, and the Willow Creek drainages and areas near Grand Lake."

"Sage grouse lek location flights- There are several areas in Grand County that we know or believe to know that there are sage grouse leks however due to access difficulty (snow or mud) we are not able to thoroughly search. We would like to visit some of these areas with the UAS to try and locate the leks. Possible locations include the Parshall Divide, Granby and the Muddy Creek drainage."

"Colorado Forest Health Project – USGS would like to fly the Raven A UAS system in support of mountain pine bark beetle tree mortality (mapping) classification work. The imagery would help sampling selection, applied throughout the various datasets the USGS uses for defining dead tree cover. Flight imagery could also help identify the extent of blow down trees in support of fuels treatments. Additional flights of fuel treatment areas could provide updates of the changes for landfire and other fuels related datasets."

"There are two current projects at UMESC where we would like to see if UAVs and remote sensing would be useful. I am conducting a study of sandhill crane habitat use in and near Horicon NWR in Southeast Wisconsin. I'd like to use UAVs to pinpoint roost and feeding locations and perhaps actually count the birds. We are also involved with surveys of pelagic birds in the Great Lakes during fall-spring. I think there may be some major hurdles with FAA for flying missions over the Great Lakes? Folks in the Geospatial Sciences branch also fly to collect aerial photography data to generate periodic land use/cover GIS. A lot of this work is done all along the Upper Mississippi River. We feel that video streaming can be handy along with the standard airplane platform, frame based aerial photography."

## End Notes

[1] http://www.doi.gov/

[2] http://www.doi.gov/bureaus/usgs.cfm

[3] http://www.usgs.gov/newsroom/article.asp?ID=1931

[4] http://amd.nbc.gov/

[5] http://www.faa.gov/about/office org/headquarters offices/avs/offices/air/hq/ engineering/uapo/

[6] http://www.faa.gov/news/testimony/news story.cfm?newsId=11599

# INDEX

## D

## M



## N



## O

## P



## Q



## R

## S



## T

## U



## V



## W